Easy Biology
STEP-BY-STEP

Master High-Frequency Concepts and Skills
for Biology Proficiency—*FAST!*

Nichole Vivion

Mc
Graw
Hill

New York   Chicago   San Francisco   Lisbon   London   Madrid   Mexico City
Milan   New Delhi   San Juan   Seoul   Singapore   Sydney   Toronto

The McGraw·Hill Companies

1 2 3 4 5 6 7 8 9 10 11 12 13 14 15 16 17   QFR/QFR   1 9 8 7 6 5 4 3 2

ISBN   978-0-07-176779-8
MHID   0-07-176779-7

e-ISBN   978-0-07-176780-4
e-MHID   0-07-176780-0

Library of Congress Control Number 2011936569

Illustrations on the following pages by Cenveo: 4, 7, 10, 13, 19 (bottom of page), 21, 27, 36, 54, 70, 78, 80, 97, 110, 122, 145, 152, 155, 163, 171, 181, 185, 187, 194, 195, 199

All other illustration and photo credits can be found on page 209, which is to be considered an extension of this copyright page.

Other titles in the series:
Easy Algebra Step-by-Step, Sandra Luna McCune and William D. Clark
Easy Chemistry Step-by-Step, Marian DeWane
Easy Mathematics Step-by-Step, William D. Clark and Sandra Luna McCune
Easy Precalculus Step-by-Step, Carolyn Wheater
Easy Writing Step-by-Step, Ann Longknife and K. D. Sullivan

McGraw-Hill products are available at special quantity discounts to use as premiums and sales promotions or for use in corporate training programs. To contact a representative, please e-mail us at bulksales@mcgraw-hill.com.

This book is printed on acid-free paper.

Contents

Preface

Easy Biology Step-by-Step is an accessible guide designed to introduce you to (or remind you of) the core concepts necessary for developing a thorough understanding of the study of life. Each chapter presents easy-to-follow descriptions of necessary information in short, targeted text with helpful supporting images.

 The Foundation checklist provides an overview of the fundamentals of the subject.

 The Step-by-Step definitions of key terms will build your understanding, including devices to help you remember their meaning.

 Vertigo panels provide problem-solving techniques and tips on how to demystify a problem.

 Misstep panels warn you of common mistakes or misconceptions.

 Following each thorough description of a new concept, a review Exercise helps to reinforce and apply new concepts.

Learning biology can seem overwhelming for a few reasons. One major factor is the breadth of information involved; biology studies everything from the organic molecules that provide the fundamental structure of living things to the complex interactions between organisms and the nonliving environment in an ecosystem. Biology doesn't skimp on depth either, for at the most fundamental level, it involves biochemistry, principles that guide the metabolism of any cell or organism.

Another issue is the sheer number of vocabulary terms involved in the discipline. While it may seem as if learning biology is like learning a new language, the more you focus on recognizing the Greek and Latin roots present in terms, the more you are building an understanding of the meaning behind a term rather than simply memorizing a definition. You will learn to predict the meaning of new terms before you ever read the definition.

Easy Biology Step-by-Step was designed with these challenging aspects of learning biology in mind. Using the written descriptions, supporting images, step-by-step problem-solving techniques, pop-up boxes, and review questions, you should be armed with the tools necessary for mastering the fundamentals of this fascinating subject.

1

The Study of Life

 ☑ Biology is the study of all life-forms and life processes. The term is derived from *bios* (meaning "life") and *logos* (meaning "study of").

☑ Biology includes a focus on both microscopic and macroscopic *levels of organization*, from cells and the smaller structures within them to tissues and organs within an individual organism and finally to the totality of ecosystems that constitute the living Earth, or biosphere.

☑ Biologists characterize something as a living thing, or *organism*, if it possesses all six of the characteristics of life to some degree.

☑ To relate the various levels of organization, it is helpful to use a thematic approach. Overarching and unifying concepts like the *relationship between structure and function* and *continuity and variation* provide a basis for connecting topics and concepts that at first may seem completely unrelated.

☑ Biology, like any natural science, relies on a set of logical steps called the *scientific method* to guide the inquiry process. The powers of observation are significant at all points of the inquiry process.

☑ Biologists use many tools to investigate nature, but none is more significant than the microscope, which enables the detection of life-forms that are too small to be seen with the naked eye.

A Thematic Perspective on Biology

If you aren't already interested in biology, you should be! Take a walk outside, and you will be surrounded by all things biological. Think about the ingenious way that plants are able to produce their own food and then thankfully share it with us. Consider how we and other animals are able to extract oxygen from the air and why we really need it in the first place. From a more selfish perspective, biology explains how your own body works and what you need to do to keep it healthy and functioning optimally.

The discipline of biology increasingly investigates many things that are invisible to the naked eye; these include some organisms like the tens of thousands of species of soil bacteria and the thousands of species of microscopic zooplankton and phytoplankton in the ocean. It also includes many things that are not living in their own right, such as viruses that imitate and parasitize life as well as the organic molecules on which all life depends—such as DNA and proteins.

How does a student of biology manage all this information? Approaching biology from a thematic perspective can be a significant tool for both organizational and memory purposes. Themes of biology can be devised and phrased in many ways, so the following six themes are suggested, overarching principles that help to connect ideas that at first seem unrelated.

The first theme is the *relationship between structure and function*; this explains that the particular form a thing takes is related to the particular task it is designed to carry out. Another theme, *continuity and variation*, is exemplified by the way in which many organisms depend on cycles to carry out life processes, as well as by the very notion of evolution and adaptation. The theme of *regulation: feedback and response* explains that organisms and the ecosystems they inhabit have mechanisms for detecting a change away from optimal conditions and can employ physiological mechanisms to bring the conditions back into normal range.

Yet another theme is *energy transfer*, which explains the variety of ways in which organisms convert one form of energy into another in their quest to carry out life. The theme of *interdependence in nature* is exemplified by the myriad relationships that different species have with one another within an ecosystem. Finally, the *process and application of science* theme demonstrates that (1) the way in which scientific inquiry is conducted is just as important as the product of that inquiry, and (2) science is often carried out with an ultimate goal of contributing positively in some way to the world around us.

Using and applying the themes of biology might seem overwhelming at this point, but once you start to think thematically, you will find many natural connections between a variety of concepts that you otherwise might not recognize. It will become easier and can become almost second nature in your approach to studying biology if you commit to it. Each of the exercises provided for additional practice will ask you to use a given theme to explain a concept, so you will have ample opportunities to do so!

Levels of Biological Organization (for All Organisms)

In addition to maintaining a thematic perspective, it is helpful to understand that the discipline of biology is structured around the notion of increasing levels of organization. The smallest level of biological organization to constitute a living thing is the cell, but there are even smaller levels that are significant in understanding how the cell and other more inclusive levels of organization function.

CHON Atoms

Atoms, the smallest naturally occurring form of matter in all substances, are predominantly in the form of carbon (C), hydrogen (H), oxygen (O), and nitrogen (N) within the cells of organisms.

Organic Molecules

Organic molecules are the stuff of life. They generally possess a backbone of carbon, a versatile atom that can form up to four bonds with neighboring atoms. These carbon skeletons can take on a variety of forms as well, from straight and branched chains to rings. The diversity in the structure of organic molecules allows for the multitude of functions required by organisms to sustain life.

Organelles

When organic molecules bind together in specific ways, they form organelles, which have specialized structures designed in the best way to carry out a given function. A typical plant cell is a chloroplast, a specialist in converting sunlight energy into food, whereas a typical animal cell is a lysosome, a small container of digestive enzymes for breaking down food molecules.

Cells

All of the organelles that are bound together within the confines of a single membrane are then part of the same cell. A *cell* is the fundamental building block of all of life; that is, an organism must be composed of at least one cell to be truly considered a living thing. If a cell contains organelles that have their own membranes, then the cell is considered *eukaryotic*. Plant, fungal, and animal cells are all eukaryotic in design. Bacterial cells, on the other hand, are considered *prokaryotic* because their organelles lack membranes.

All organisms, from the simplest bacteria to the most complex plants and animals, share the cellular level of biological organization. A cell membrane (or *plasma membrane*)—made up of an array of specific organic molecules rich in carbon, hydrogen, oxygen, and nitrogen atoms—provides an outer boundary that is able to regulate what enters and leaves the cell. The internal *aqueous cytoplasm* gives the cell an appropriate medium within which to carry out the chemical reactions that are part of its metabolism.

Deoxyribonucleic acid (DNA), whether found in the cytoplasm of a prokaryote or housed within the membrane-bound nucleus of a eukaryote, is another essential organic molecule. It provides the instructions for building

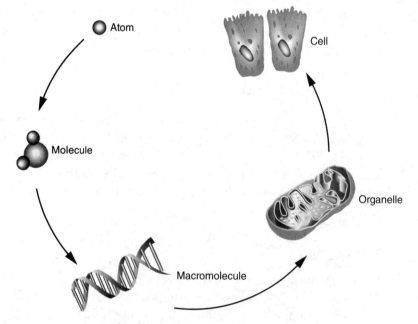

Atom

Molecule

Macromolecule

Cell

Organelle

Levels of organization within all organisms

all of the necessary proteins that assist in the cell's metabolism and make up some structural components of the cell itself. A very important and plentiful organelle, the *ribosome*, uses the instructions from DNA to build those proteins.

 The most important thing to understand about the biological levels of organization is that an organism must possess a lower (less inclusive) level of organization in order to possess a higher (more inclusive) level. For example, for a plant to have a specialized organ like a flower, that organ must have true tissues made up of specialized cells working together for a common purpose. The cells in those tissues will look different and function differently from those of a different tissue within the individual plant. Without this fundamental cell structure, an organism would not be able to carry out all necessary life functions and would not possess all of the characteristics of life (discussed later in this chapter).

Exercise 1.1

If you can successfully work through the following questions, you should feel confident about being able to go on to the next section.

1. An atom of carbon or hydrogen is to an organic molecule as _____ is (are) to a cell.
   a. an organic macromolecule
   b. an atom of oxygen or nitrogen
   c. an organelle
   d. all of the above

2. Which of the following organisms would *not* possess tissues?
   a. A bacterium such as *E. coli*
   b. A plant such as a pine tree
   c. An animal such as an earthworm
   d. A fungus such as a mushroom

3. Use the theme of the relationship between structure and function to describe the concept of a cell.

   _____

   _____

   _____

Levels of Biological Organization (for Multicellular Organisms)

Now that you have mastered the idea of increasing levels of organization within unicellular organisms, it should be easy to understand how the same applies to multicellular organisms. The levels of organization simply become more complex and require additional levels of coordination and integration to keep the organism functioning.

Tissues

If an organism is *multicellular*, then it often has collections of different, specialized cells working together for a common purpose. This constitutes a *tissue*. A characteristic animal tissue type is muscle, which facilitates movement through its contractile properties. A typical plant tissue is vascular, meaning it is designed to transport water and food throughout the various parts of the plant body.

Organs

When different tissues work together for a common purpose within a specific region of an organism, the organ level of biological organization has been reached. Typical plant organs include the leaf and the root, whereas typical animal organs include the gut and the heart.

Organ Systems

When different organs coordinate efforts to achieve a desired outcome for an organism, they constitute an *organ system*. In plants, organ systems include the root system (organs below ground) and the shoot system (organs above ground). In animals, the organ systems are typically much more numerous. Animal organ systems include the digestive and circulatory.

Organism

An organism is any living thing. Some of the organisms with the simplest structures are *unicellular*, composed of only one cell and thus invisible to the naked eye. Examples are bacterial species like *E. coli* and *Salmonella*. Many others are *multicellular*, composed of many cells. The typical human body is

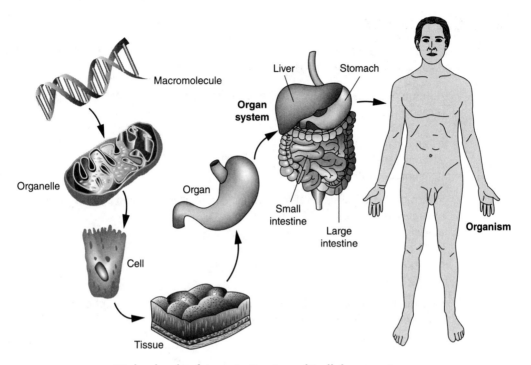

Macromolecule

Organelle

Organ system

Liver

Stomach

Organ

Cell

Small intestine

Large intestine

Organism

Tissue

Higher levels of organization in multicellular organisms

composed of trillions of its own cells and, somewhat shockingly, is also populated with even more bacterial cells inside certain organs and on body surfaces.

As you can now understand, organisms are greatly dependent on appropriate and functional organization. A cell may have a specialized function by being composed of a high concentration of a particular organelle. A population may be out of balance if the number of individuals of one biological sex differs vastly from the number of those of the other.

Another interesting concept that is illustrated by the biological levels of organization is that of emergent properties. An *emergent property* is one that does not exist at one particular level but then emerges as a significant characteristic at the next, higher level. For example, neither an organelle within a heart muscle cell nor the muscle cell itself has a heart rate, but the collective organ of which they are a part—the heart—does. Emergent properties account for the synergetic notion that the whole is greater than the sum of its parts.

Exercise 1.2

If you can successfully work through the following questions, you should feel confident about being able to go on to the next section.

1. A group of specialized cells working together to carry out a specific function is called _____.
   a. a tissue
   b. an organ
   c. an organelle
   d. an organ system

2. Which of the following levels of biological organization would possess more emergent properties than the others?
   a. Cell
   b. Organ system
   c. Organ
   d. Organism

3. Use the theme of continuity and variation to describe the concept of organs and organ systems in a complex, multicellular organism.

   _____

   _____

   _____

The Six Characteristics of Life

For something to be considered a living thing, and thus relevant to biological study, it must have all of the following six characteristics of life. Even if it had most of the characteristics and lacked only one or two, it would still not technically qualify as an organism. This helps to distinguish bacteria from viruses, two familiar things that can make us sick. There is a big difference between them: bacteria are unicellular organisms that meet all six characteristics of life, whereas viruses are not composed of even a single cell and thus cannot meet the strict definition of life.

Cells and Organization

The lowest level of biological organization known to be able to support life is the cell. A cell's contents are contained by its own plasma membrane, and its traits and activities are directed by its genetic code, which is stored in the form of DNA. Organelles specialize in structure and function and then coordinate activities within the cell to support all of life's processes.

Presence of Metabolism

Metabolism is the sum of all the chemical reactions that take place within a cell or an organism. These chemical reactions involve reactions of two main types: those that are *catabolic* are responsible for breaking larger substances down into smaller ones, and those that are *anabolic* build up larger substances from smaller ones.

Ability to Reproduce

Reproduction, the creation of new organisms from existing ones, occurs in one of two main forms, depending on the number of parents involved. When a single parent produces an offspring, *asexual reproduction* occurs. The offspring is typically a genetic clone of the parent, because the asexual parent passes on an exact copy of its *genome*, a set of all of its genes, to its offspring. Each *gene* contains a code for making a protein that helps the organism express a trait. *Sexual reproduction* involves two parents, each of which passes along only half of its genetic material to its offspring through specialized reproductive cells. Development is typical in more complex, multicellular organisms when a juvenile form changes morphologically over time to achieve the adult form.

Ability to Grow

Organisms achieve growth in one of two primary ways: *cell enlargement* and *cell division*. If an individual cell increases its size by increasing its surface area and volume, then cell enlargement has occurred. However, if a cell splits to create two offspring cells, then cell division has occurred. If an organism is unicellular, then it can only undergo cell enlargement; if it divides, then it is actually reproducing. Multicellular organisms experience some degree of both cell enlargement and cell division in order to achieve overall growth.

Maintenance of Homeostasis

Homeostasis is an absolutely critical survival characteristic for organisms in ever-variable and unpredictable environments. It involves the mechanisms by which cells (and thereby the higher organizational levels, like organs and

The six characteristics of life evident within a pond ecosystem

organisms) can maintain stable internal conditions even when the external environment in which they are placed is variable.

Detection of and Response to Stimuli

Cells and organisms must have some means of picking up on important cues in the environment that might affect their well-being and then reacting appropriately. For example, many organisms have an ability to detect levels of light (through photoreceptors in eyes and photosynthetic organelles called *chloroplasts* in plant leaf cells) and may adjust position to interact optimally with the given light levels.

It is easy to confuse the six characteristics of life with the six themes of biology. It is important to remember that something must possess all six characteristics of life to be classified as an organism. These characteristics can thus be used to explain why a bacterial cell is considered alive, whereas a virus is not. The themes of biology are six overarching principles that help us relate different concepts across all units within the discipline.

Exercise 1.3

If you can successfully work through the following questions, you should feel confident about being able to go on to the next section.

1. When an animal cell breaks down sugar to release energy in the process known as cellular respiration (to be discussed in detail in Chapter 4), the cell is directly carrying out which characteristic of life?
    a. Growth
    b. Metabolism
    c. Homeostasis
    d. Response to stimuli

2. When a plant adds more cells to its body, it is experiencing _____; when it releases seeds in a fruit, it is involved in _____.
    a. homeostasis; reproduction
    b. reproduction; growth
    c. growth; reproduction
    d. development; homeostasis

3. Use the theme of energy transfer to explain the concept of metabolism as an essential characteristic of life.

    _____

    _____

    _____

The Scientific Method

Scientists of all types rely on a logical, organized approach to investigate the physical and biological world. That approach, the scientific method, is important in that it encourages objectivity and discourages bias. It also allows other scientists to replicate the procedure and verify (or refute) the results.

Making Observations

Observations involve the use of the senses to perceive a natural phenomenon. Humans tend to rely heavily on sight to observe visual stimuli, but stimuli perceived through hearing, touch, taste, and smell can also be significant, depending on the conditions.

Forming a Hypothesis

A hypothesis is an educated guess about the outcome of a natural phenomenon or event based on prior knowledge and observations. It may include a specific prediction about the outcome should the hypothesis be true.

Conducting an Experiment

Data must be collected during an experiment to use as evidence for the support or rejection of the hypothesis. Data can be *qualitative* (described in words) or *quantitative* (described in numbers). A *controlled experiment* is the standard in science, as it provides a way to control for all factors except the one idea being tested. In a controlled experiment, the researchers separate test subjects into two groups, the *control group* and the *experimental group*. The two groups should be identical in all ways, except that the experimental group receives the experimental treatment under evaluation and the control group does not.

The treatment applied to the experimental group and withheld from the control group describes the *independent variable* that is specifically controlled by the researchers and will potentially cause some change in the group to which it is applied. Much of the data collected during the experiment will involve the *dependent variable*, or *outcome variable*. This is the predicted outcome of the independent variable that should be observed in the experimental group and not the control group.

Analyzing Results and Drawing Conclusions

The data are analyzed, often through the use of statistics, graphs, and models. The effect the independent variable had on the dependent variable is examined by comparing the results for the control group with those of the experimental group. Any significant variation is assumed to be due to the only difference, the treatment of the independent variable in the experimental group.

Contributing to a Body of Scientific Knowledge

The methodology employed and results obtained in any experiment can be shared clearly and consistently with the scientific community through research articles in scientific, peer-reviewed journals. If the results can't be replicated under identical conditions, then the conclusions may be called into question once again.

The concepts of independent and dependent variables are easily confused. It may help to consider the independent variable as the condition manipulated by the biologist conducting the experiment. Similarly, the dependent variable should be considered the outcome of the experiment; it thus *depends* on whether the independent variable was applied to a group.

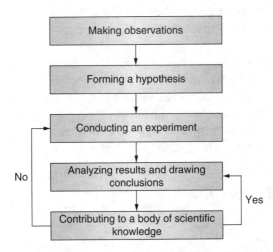

A flowchart of the steps within the scientific method

The difference between experimental and control groups can be perplexing. Remember that of the two, the experimental group receives the treatment of interest ("+ independent variable"), while the control group does not receive the treatment of interest ("− independent variable"). The control group thus serves as the point of comparison and allows researchers to determine if the independent variable was responsible for any difference (dependent variable) observed between the two groups.

Although the scientific method is often described as comprising the series of steps presented here, there are really a vast number of different and equally valid approaches to scientific inquiry. The power of observation and the ability to pay close attention to detail are valuable skills in carrying out scientific experiments. For the results of a particular experiment to be considered valid, others must be able to replicate the procedure and produce the same results. If they can, the conclusions may then be adopted by the discipline and the larger scientific community.

For science to be collaborative on the global level, clear and consistent communication is important. One way this is demonstrated is through reliance on the *peer review process* before scientific research is published in the literature. Experts in the field must agree that the process is valid and that there is no apparent bias inherent in the experimental design that might have affected the results. Another demonstration of the significance of communication in science is through the use of the metric system for data collection. The units of the metric system are based on powers of 10, so conversions and calculations are relatively easy.

 Theory is one of the most misunderstood and commonly misused terms in all science. (Usually, one should use *hypothesis*.) While in common conversation, it may mean "just an idea," the term holds much more weight in science. In fact, it should be equated in principle to the term *law*. The theory of evolution by means of natural selection and the cell theory are both large, overarching principles, each supported by a massive body of evidence.

The Microscope

A microscope is an essential tool for many biologists. The lenses within a microscope allow the biologist to magnify an object, or enlarge its apparent size, so it can be seen by the human eye. Adjusting other parts of the microscope—the coarse focus and the fine focus—helps to refine the resolution, or clarity, with which the enlarged specimen can be seen.

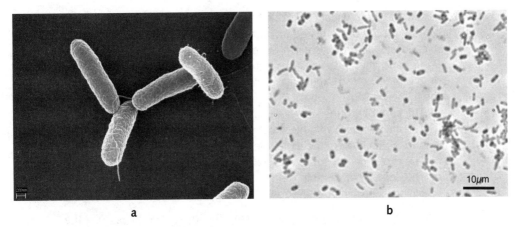

a   b

Salmonella **bacteria under (a) an electron microscope and (b) a light microscope**

For a long time, only light microscopes were available to biologists. Light microscopes, particularly *compound light microscopes* with a set of two lenses, are still of great use today, for they are the most powerful types of microscopes and allow us to see organisms while they are still living; these microscopes are relatively inexpensive and easy to use. Their more recent counterpart, the *electron microscope*, allows us to peer inside cells and see the details of cell and organelle surfaces. The downside of electron microscopes is that organisms must be dead, the equipment is very expensive and physically large, and users require specialized training.

Exercise 1.4

If you can successfully work through the following questions, you should feel confident about being able to go on to the next section.

1. Which of the following steps in the scientific method is usually also involved in all the other steps?
   a. Forming a hypothesis
   b. Making observations
   c. Analyzing data
   d. Experimenting

2. Before it is made widely available through medical prescriptions, a pharmaceutical drug is being tested on a group of humans who regularly suffer from migraines. One group is given the drug, while the other is given a placebo (a sugar pill known to cause no significant effect). The group that is given the drug is predicted to experience a reduction in migraine headaches. Which of the following is true?
   a. The control group receives the drug.
   b. The independent variable should be the drug applied to the control group and withheld from the experimental group.
   c. The experimental group receives the placebo.
   d. The dependent variable is the number of migraine headaches experienced by individuals in both groups.

3. Use the theme of the process and application of science to describe the significance of the microscope to the study of biology.

   _____

   _____

   _____

2

The Chemistry of Life

☑ Organisms are made up of at least one cell, and cells are made up of organic molecules filled with carbon, hydrogen, oxygen, and nitrogen atoms.

☑ Atoms make up all of the matter around us, whether *organic*—part of an organism—or *inorganic*—part of nonliving matter. *Matter* is defined as something that has mass and takes up space.

☑ Chemistry studies all matter, whether organic or inorganic; biology focuses mainly on organic matter as well as the chemistry of water, an essential inorganic molecule for life.

☑ *Organic macromolecules* like proteins and carbohydrates are essential in making up the physical structures of organisms as well as carrying out and controlling their metabolic functions.

Basic Chemistry

It should start to become obvious how essential chemistry is to the study of life. Understanding the structure of individual atoms and the ways in which atoms interact to form compounds and molecules is fundamental in appreciating the metabolism of organisms.

Subatomic Particles

Subatomic particles are the smallest indivisible particles in nature. They include the *proton* (positively charged), the *electron* (negatively charged), and the *neutron* (neutral). The number of protons uniquely defines a particular type of atom and establishes its basic chemical behavior; if the proton number changes, the chemical substance changes. Neutrons can vary in number within atoms of the same type of element; the resultant atoms are called *isotopes*. Electrons are involved in chemical bonding, or linking different atoms together (described in detail later in this chapter). They can move from atom to atom more easily than other atomic particles, but their mass is so tiny that it is negligible relative to the much heavier protons and neutrons.

Atoms

An *atom* is the smallest unit of substance in nature and is considered the building block of all matter. Every atom has a very small, central core called the *nucleus*. Because the protons and neutrons are housed in the nucleus, it possesses an overall positive charge. In the vast space outside the nucleus, any electrons contained within the atom orbit at various energy levels, or certain regions at characteristic distances from the nucleus. The space outside the nucleus is thus negatively charged.

Atoms are considered neutral because they have equal numbers of protons and electrons. When a pure substance consists of only one type of atom, that substance is considered an *element*. All known chemical elements are arranged in a helpful resource called the *periodic table*.

Compounds and Molecules

When atoms of two or more elements are bonded into a more complex structure, a *compound* is created. Often a compound is formed between two different types of atoms that *ionize* before bonding. This means that, as they

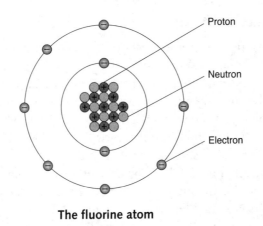

The fluorine atom

approach each other, one or more electrons jumps from the outermost energy level, or *valence level*, of one atom to the valence level of the other. The one that lost one or more electrons becomes positive and is now called a *cation*. The atom that gained one or more electrons becomes negative and is referred to as an *anion*. Now that the two atoms are oppositely charged, an *ionic bond* readily forms between the two. While very strong, an ionic bond can easily break if placed near other charged particles.

In other cases, two or more atoms (of the same or different elements) can overlap their valence levels and, by doing so, share the electrons they possess at that particular level. This creates a *covalent bond* between the atoms and creates a more complex structure called a *molecule*. While the bond energy between atoms in a covalent bond might not be as strong as that in an ionic bond, the covalent bond is much more durable. Other charged substances do not disrupt this type of bond.

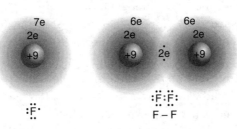

The covalent bonds in fluorine gas (F_2)

Mixtures and Solutions

When two or more pure substances are combined into a larger substance, a *mixture* is created. When one substance within a mixture dissolves evenly in the other, a *solution* is present. A solution comprises two major parts: the *solute* (the substance being dissolved) and the *solvent* (the substance doing the dissolving). When the solvent is water, as it is in most biologically relevant solutions, the solution is referred to as *aqueous*. In aqueous solutions, the solute is typically a substance that is fully charged, such as an ion, or partially charged (called *polar*), like sugar or other water molecules. Polar molecules result from the unequal sharing of electrons between the atoms in a covalent bond. The atom that has a relatively high electronegativity typically has more electrons near it, while the relatively less electronegative atom has fewer electrons in its vicinity. This leaves a partially negative charge on the former atom and a partially positive charge on the latter.

Although we are primarily concerned with biology, it is important to remember that organisms behave according to the laws of chemistry as well. This is because living things are composed of matter (that is, they have mass— a measurable quantity of matter—and take up space) just as nonliving things are. That means everything you can see, beyond the light you are using to see, is made of matter. So are many things that you cannot see, including the mixture of various gases that make up the Earth's atmosphere.

The Three States of Matter

We are familiar with three states of matter on Earth: *solid*, *liquid*, and *gas*. Consider the ever-important water molecule; it commonly exists in all three forms because its freezing and melting points lie within the usual ambient temperature range we experience. Solid water (ice) can change state into liquid water if exposed to enough energy. Scientifically, *energy* is defined as the ability to do work. The work in this scenario involves changing the way the water molecules are behaving.

In a solid, the water molecules are vibrating in place and are confined to a specific space and shape. When energy is transferred to ice, the molecules absorb some of that energy and begin to move around more as they become a liquid. Liquids are less confined than solids, so liquid water will take the shape of any container in which it is placed, although it still has a defined volume as it did when it was ice. If even more energy is transferred to liquid water, the molecules can eventually change state again and become gaseous. In a gas, the molecules move freely and randomly, so much so that they will always expand to fit their container and therefore have no definite volume.

Energy and Metabolic Reactions

The concept of energy becomes significant when we consider an organism's metabolism. Recall that metabolism includes all of the chemical reactions that take place within a cell or an organism, including those that break down polymers and those that build them up. In any chemical reaction, energy is transferred as certain bonds between atoms are broken and new ones are formed. The inputs of any chemical reaction are called *reactants*, while the outputs are called *products*. Consider the following reaction:

$$2H_2 + O_2 \rightarrow 2H_2O$$

This reaction describes how hydrogen gas (H_2) and oxygen gas (O_2) can combine to produce water (H_2O). To adhere to the *law of conservation of matter*, two molecules of H_2 and one molecule of O_2 constitute the reactants, while two molecules of H_2O make up the products. You may be wondering why, in times of drought, we cannot simply make water using chemistry. While the reaction is chemically possible, it is not easy to achieve due to the huge amount of energy required. This is because the reaction is *endothermic*, meaning the reactants must also be exposed to a certain amount of energy in order to form products.

Other reactions are *exothermic*, meaning that product formation is associated with a release of energy. Within a cell, an exothermic reaction is often linked to an endothermic one in an energy-efficient process called *coupling*. That way, the energy released in one metabolic reaction can be used directly to drive the energy needed in another. Whether endothermic or exothermic, every reaction has an *activation energy*, the high energy state that must be surmounted to create products from reactants.

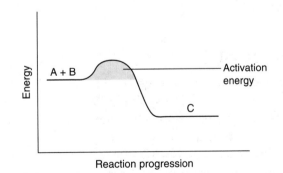

Activation energy of an exothermic reaction

Exercise 2.1

If you can successfully work through the following questions, you should feel confident about being able to go on to the next section.

1. If an individual carbon atom possesses six protons in its nucleus, then how many electrons should be orbiting outside?
    a. One
    b. Four
    c. Six
    d. Eight

2. If the same carbon atom were to share four electrons with a neighboring carbon atom, what type of chemical bond would be formed?
    a. A saturated bond
    b. An ionic bond
    c. A covalent bond
    d. A molecular bond

3. Use the theme of energy transfer to describe the concept of the three states of matter.

    _____

    _____

    _____

Carbon Chemistry

Of any of the chemical elements on the periodic table, carbon is undoubtedly the most significant to life. Although a large proportion of the mass of an individual organism can be attributed to internal water, the atom that makes up the physical matrix for organic molecules and cells, and thus contributes to its *biomass*, is carbon.

Carbon Atom

Carbon is considered *the* organic element. It is exceedingly versatile due to its four valence electrons. *Valence electrons* are those in the outermost energy level, and because bonds form only when outer electrons are shared or transferred, they are the only electrons potentially available for the formation of chemical bonds with another atom. With four valence electrons, a carbon atom can form up to four different bonds with neighboring atoms and can thus contribute to organic molecules of diverse structures and functions.

Organic Molecule

When a carbon atom bonds to one or more other carbon atoms to produce a structure with a carbon backbone, an organic molecule is created. Some organic molecules are considered *hydrocarbons* because they are composed of a carbon backbone surrounded only by hydrogen atoms. These molecules are completely uncharged and are referred to as *nonpolar*. This lack of any charge makes nonpolar substances insoluble in water, so they are also called *hydrophobic*.

Many organic molecules are not hydrocarbons; they have had a hydrogen atom removed and replaced with a functional group—a specific set of atoms that is responsible for predictable reactivity in any molecule of which it is a part.

Having four valence electrons also means that it is easy for a carbon atom to form different types of bonds. Not only does carbon readily form single bonds with neighboring carbons (in which one pair of electrons is shared), but it can also form a *double bond* (in which two pairs of electrons are shared) or a *triple bond* (in which three pairs of electrons are shared). This acts to increase the structural complexity and diversity of organic molecules, thereby permitting a vast array of functions.

Monomers

Monomers are the fundamental building blocks of organic macromolecules, those on which all life-forms are based. *Monomers*, which are organic molecules, are linked together through covalent bonds to form larger *polymers*. If a monomer is broken down, then it loses the structural and functional properties of the larger macromolecule type to which it belongs.

Polymers

Polymers are the product of many like monomers linked together through covalent bonds. Technically, two monomers bonded together constitute a *dimer*, while three or more linked monomers form a polymer. These chemical bonds within a polymer can be broken down to release the individual monomer building blocks. Thus, a complex polymer within some food molecules that we eat can be digested into simpler parts. These monomers can then be absorbed into the bloodstream and used wherever needed by the body's cells.

Organic Macromolecules

Organic macromolecules are the larger categories into which the basic polymers fit. The four basic macromolecules that organisms depend on are *proteins, carbohydrates, lipids,* and *nucleic acids* (discussed in detail in the next section). Each organic macromolecule is characterized by a specific structure and therefore is associated with a given set of typical functions that helps the cell and/or organism carry out life processes.

The concept of monomers and polymers can seem confusing at first. It is helpful to think of monomers like letters in the English alphabet and polymers as the words they can form. Just as different monomers within a specific macromolecule type can be arranged into polymers of varying sequences and lengths, the 26 letters can be arranged in so many ways as to account for a seemingly endless supply of words.

Cellular Metabolism

The concept of monomers and polymers is essential in understanding cellular metabolism. As stated previously, anabolism involves building up larger molecules from smaller ones, while catabolism involves the reverse mechanism of breaking down a large polymer into its component monomer parts. How exactly can a cell accomplish this chemically? Two specific types of reactions are employed: *hydrolysis* and *dehydration synthesis* (also called *condensation reaction*).

When polymers are broken down through a hydrolysis reaction, a water molecule is invested and atoms from it are used to force apart the existing bond. Digestion thus involves hydrolysis reactions and contributes to catabolism.

When individual monomers are linked together to form polymers, a dehydration synthesis is employed. A hydroxyl group ($-OH$) is removed from one monomer, and a hydrogen atom ($-H$) is removed from the other; these pieces are linked together to produce water. The two monomers then link together where they lost atoms in the dehydration process, and a polymer is produced. This contributes to anabolism of the cell and occurs when a cell produces new enzymes or grows more muscle fibers.

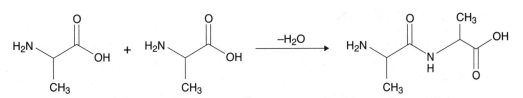

Dehydration synthesis between amino acids to form a dipeptide

 Some molecules contain carbon but are not strictly organic because they lack multiple carbon atoms. For example, carbon dioxide (CO_2) is inorganic but is essential to organisms in providing the basis by which most organic molecules are synthesized through the process of photosynthesis. Methane (CH_4) is on the boundary between inorganic and organic. It, too, can act as a building block for organic molecules and is produced and released by many organisms as part of their metabolism.

 ## Exercise 2.2

If you can successfully work through the following questions, you should feel confident about being able to go on to the next section.

1. An organic molecule will always contain which of the following?
   a. Nitrogen
   b. Oxygen
   c. Carbon
   d. All of the above

2. The building of glucose molecules from carbon dioxide by plants involves
   _____.

   a. anabolism
   b. catabolism
   c. a dehydration synthesis reaction
   d. both a and c

3. Use the theme of continuity and variation to describe the concept of monomers and polymers in organic molecules.

   _____

   _____

   _____

Organic Macromolecules

Now that you realize the versatility of the carbon atom, it is important to appreciate the complexity and diversity of the different organic molecules in which carbon atoms are found. These organic molecules provide a physical means of connecting the metabolisms of all organisms on Earth.

Macromolecules of Life

The macromolecules of life can be thought of as the three basic categories of food from which we get caloric energy through our diet (namely, proteins, carbohydrates, and lipids—a more inclusive category than fats). They are also the molecules that store and process genetic instructions (specifically, nucleic acids like DNA and RNA [ribonucleic acid]).

Proteins

The monomer form of a protein is called an *amino acid*; two amino acids bonded together make a *dipeptide*, while three or more constitute a *polypeptide*. Each amino acid is composed of a central carbon atom bonded to four things: an *amino group* ($-NH_2$), a *carboxyl group* ($-COOH$), a *hydrogen atom* ($-H$), and a variable collection of atoms called the *R group*. Each of the 20 amino acids found in proteins has a unique R group that distinguishes it from the others.

Proteins provide a multitude of functions. Many proteins are structural, making up a significant portion of the physical foundation of a cell or an organism. These include the long, fibrous proteins that make up the cytoskeleton present as internal scaffolding in eukaryotic cells and the overlapping proteins that allow skeletal muscles to contract and relax. Other proteins are considered functional; these primarily include biological catalysts called *enzymes* that act to speed up chemical reactions so these reactions can occur within a suitable time frame to meet the metabolic needs of the cell or organism. The presence or absence of an enzyme can thus act to regulate metabolism; if the enzyme is present, the reaction will occur, whereas its absence means the reaction will not take place.

Carbohydrates

The monomer form of a carbohydrate is called a *monosaccharide*; two monosaccharides bonded together make a *disaccharide*, and three or more con-

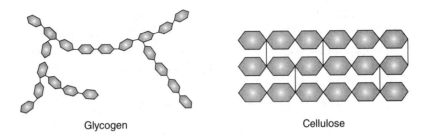

Glycogen   Cellulose

(Note that each ring is a monosaccharide)

Glucose monomers linked into glycogen and cellulose, two common polysaccharides

stitute a *polysaccharide*. Each monosaccharide is made up of only carbon, hydrogen, and oxygen in the specific ratio of $(CH_2O)_n$.

In our diet, *simple sugars* like sucrose (table sugar) and lactose (milk sugar) are typically disaccharides associated with quick energy, because only one bond needs to be broken before each monosaccharide unit can be used for energy. Complex carbohydrates like *starches* consist of long, branched chains of monosaccharides linked together with many bonds that need to be broken, so they tend to provide longer-lasting energy than simple sugars.

Lipids

Lipids differ from the rest of the macromolecules in that they don't have a typical monomer and polymer form. Because they are not water soluble, their chemical behavior is different. Lipids actually don't behave according to the typical "monomer + monomer = dimer" rule. They instead have a specific structure that can vary quite markedly depending on the type of lipid.

For example, a major type is the *triglyceride*, the lipid polymer typically found in the foods we eat. The triglyceride is made up of a glycerol molecule and three fatty acid chains. The glycerol, a nonlipid molecule on its own, provides a three-carbon backbone to which each of the fatty acid chains attaches. These fatty acid chains are really just long series of carbon atoms surrounded by hydrogen atoms. If there are only single bonds connecting carbons in the chain, then the chain is described as *saturated*; if one or more double bonds is present, then the chain is described as *unsaturated*.

In other types of lipids, a different combination of component parts is involved. For example, a *phospholipid*, the main component of cell membranes, is made up of a glycerol backbone and two fatty acid chains. The third carbon on the glycerol backbone, however, is attached to a phosphate group, a negatively charged group of oxygen atoms surrounding a phosphate atom. The presence of the charged phosphate group drastically changes the function of the molecule.

The other major groups of lipids—waxes and steroids—have their own unique recipe. A *wax* consists of one long fatty acid chain connected to an alcohol group ($-OH$), making that end charged. Waxes are useful to organisms in providing a watertight boundary, and the fatty acid chain means that we can use them in a candle for long-lasting light and heat energy when combusted (burned in the presence of oxygen). *Steroids* have the most unique structure of all the lipid groups, because no fatty acid chain is present. Instead, four fused carbon-based rings are present on the molecular level. This divergent structure allows steroids to behave very differently; they include the molecule *cholesterol* and also often act as *hormones* (chemical messengers) in plants and animals.

The steroid type of lipid is often misunderstood. We often immediately associate the term *steroids* with illicit bodybuilding substances abused by some professional athletes. These substances are often lipid molecules that mimic natural steroids found in the body, but there are many other types of useful and essential steroids that are not associated with building muscle mass. The primary sex hormones in humans, estrogen and testosterone, are two examples of such steroids. They are responsible for triggering the development of secondary sex characteristics that takes place during puberty and also for the maturation of sex cells like eggs and sperm.

Nucleic Acids

Nucleic acids differ from the rest of the biological macromolecules in that they provide negligible caloric energy in our diet (which is why they are not listed on a nutritional label), but they are just as significant. Nucleic acids come in two basic polymer types, DNA and RNA. Each type is composed of at least one chain of nucleotides, the monomer unit. A *nucleotide* is a complex structure composed of a pentose (five-carbon) sugar, a phosphate group, and one of four *nitrogenous bases*: adenine (A), cytosine (C), thymine (T), and guanine (G). When nucleotides are covalently bonded to other nucleotides, a polynucleotide is created. Usually, at least thousands of nucleotides are bonded together to make up a meaningful polynucleotide from the perspective of an organism.

Nucleic acids hold the key to the genetic code. DNA stores the instructions for building proteins in the specific sequence of As, Cs, Ts, and Gs present within a stretch of the molecule known as a *gene*. RNA makes a copy of the gene when needed and then carries these instructions in its own sequence of nucleotides (called an *RNA transcript*) to a ribosome. The *ribosome*, an essential structure within all known cells, is structured such that it can read the transcript and build a specific protein by linking together certain amino acids.

One important organic molecule that has yet to be discussed is *adenosine triphosphate* (ATP). It is considered the energy currency for all cells, meaning that when a cell needs to do work to carry out a life function, it most often uses ATP to do so. As you might imagine, the structure of ATP is related to the way in which it can be used as cellular energy. ATP is composed of the specific RNA nucleotide adenine attached covalently to three phosphate groups. When the bond holding the terminal phosphate group is broken, the energy that was holding the functional group to the rest of the molecule is released and can be used to do cellular work.

You may be wondering how to distinguish between organic macromolecules. Think about the CHON atoms! Carbohydrates (as their name implies) contain carbon and the components of water (hydrogen and oxygen). Thus, carbs possess C, H, and O atoms but never any N. The same is true for lipids, but they possess many more C and H atoms and relatively few O atoms. Proteins, on the other hand, are rich in N atoms due to the presence of the amine group in every monomer. They also are the only macromolecule type ever to contain sulfur, as one main amino acid possesses a sulfur atom in its R group and can use it to form special bonds that are unique to proteins. Finally, nucleic acids always have C, H, O, and N atoms like proteins, but they never contain sulfur and always contain phosphorus (due to a phosphate group in each monomer).

Exercise 2.3

If you can successfully work through the following questions, you should feel confident about being able to go on to the next section.

1. When thousands of amino acids are bonded together, what polymer form is created?
   a. A polypeptide
   b. A polysaccharide
   c. A triglyceride
   d. A polynucleotide

2. What type of macromolecule is found within the plasma membranes of all cells?
   a. A triglyceride
   b. A wax
   c. A steroid
   d. A phospholipid

3. Use the theme of the relationship between structure and function to describe the concept of carbohydrates and lipids as forms of energy storage.

   _____

   _____

   _____

3

Cell Structures and Functions

☑ Cells are the basic organizational units of all life, from the simplest bacteria to relatively complex plants and animals.

☑ An individual cell must be structured in such a way that it is capable of carrying out all of its necessary life functions.

☑ Some cells are relatively simple, containing just one membrane that separates the internal contents from the outside world; they are called prokaryotes. Other cells have internal membranes that allow increased specialization in structure and function; they are called eukaryotes.

☑ Plant cells and animal cells are structurally and functionally distinct due to the presence or absence of several organelles.

☑ Individual cells within a complex, multicellular organism might specialize in structure and function, or they may differentiate. This requires communication and coordination between cells for the overall organism to function appropriately.

Cell Diversity

Cells have evolved into many distinct forms in order to carry out the multitude of specialized life functions in often very diverse environments. At the heart of this evolution is the relationship between structure and function, illustrating that the size, shape, and organizational complexity of an organism are indicative of the processes it performs.

Size and Shape

Cells come in a wide variety of sizes and shapes, depending largely on their function and environment. The only cell present in a unicellular organism is structured to best fit its environment. If it is in an aqueous cell, it may possess a long, tail-like structure called a *flagellum* to help propel it around. Cells must maintain an appropriate surface-area-to-volume ratio (SA/V) to be able to satisfactorily carry out all life processes.

Diversity in the structure of human blood cells

Degree of Internal Specialization

Some cells are relatively simple in structure and don't contain separate, internal compartments that carry out specialized tasks. They are called prokaryotic cells, or *prokaryotes*. Other cells, known as *eukaryotes*, are more complex, containing a variety of specialized organelles that coordinate together for overall functioning of the cell. A unicellular organism, whether prokaryotic or eukaryotic, must be composed of a generalist cell that can meet all the demands of life.

A multicellular organism, on the other hand, can be made up of highly specialized cells that may work together to constitute a tissue or other higher levels of biological organization. All multicellular organisms are eukaryotic, but the degree of structural specialization depends on the type of organism. Plants and animals tend to show the highest degree of structural complexity.

Intracellular Organization

At one point in our evolutionary history, all organisms on Earth were unicellular. Eventually, some came together in small groups called *colonies*. Individual cells within a colony were still considered individual organisms, but they benefited from the sum activity of the whole. In some instances, individuals within a given colony became so interdependent (associated with significant genetic change) that they evolved into a new form of life—a true multicellular organism.

In trying to keep track of the difference between prokaryotic and eukaryotic, it helps to understand the roots of the words. The key root common to both terms, *karyo*, refers to a "core," and in this case, more specifically, a nucleus. Thus, prokaryotes (*pro* means "before") are cells that existed before the nucleus evolved, while eukaryotes (*eu* means "true") possess a true nucleus.

As already mentioned, maintenance of an appropriate SA/V is crucial to a cell's functioning. If a cell gets too large, it may have too much volume and not enough surface area along its plasma membrane to efficiently regulate the comings and goings of materials into and out of the cell. This limitation on a cell's size is best understood when assuming a cell has a spherical shape. The surface area of a sphere is defined by the formula $SA = 4\pi r^2$, while the formula for volume is defined as $V = 4/3\pi r^3$ (where r = the radius of the sphere). This means that as the cell grows and the radius increases, the volume increases at a much faster rate than the surface area.

Exercise 3.1

If you can successfully work through the following questions, you should feel confident about being able to move on to the next section.

1. Which type of cell would you expect to have the highest degree of structural complexity?
   a. A plant cell
   b. A unicellular, eukaryotic cell
   c. A bacterial cell
   d. A prokaryotic cell

2. If a cell's radius increases from 1 micrometer to 10 micrometers, by what factor does the surface area increase? By what factor does the volume increase?
   a. By a factor of 100; by a factor of 1,000
   b. By a factor of 1; by a factor of 10
   c. By a factor of 400; by a factor of 4,000/3
   d. By a factor of 1,000; by a factor of 100

3. Use the theme of interdependence in nature to describe the concept of a colony in the evolution of multicellularity.

   _____

   _____

   _____

The Prokaryotic Cell

This type of cell possesses a membrane to protect it and to regulate the transport of materials into and out of the cell; it also possesses an internal, jellylike fluid and suspended organelles and molecules collectively called the *cytoplasm*. The major internal structures include ribosomes; organelles to help build proteins; and DNA, the molecule that stores the genetic instructions required for that cell to carry out life.

Plasma Membrane

The plasma membrane (or cell membrane) is an essential component of all cells. It is made up primarily of phospholipids that are specifically arranged into a *bilayer*. Proteins of various sorts are scattered throughout the membrane; some only stud the surface, while others are fully embedded within both layers of phospholipids. Because the membrane is the surface that directly separates the internal contents of the cell and the external environ-ment, it is important for protection and regulation. (The plasma membrane will be discussed in detail in Chapter 4.)

Cytoplasm

The cytoplasm is the aqueous solution within which chemical reactions that are part of the cell's metabolism may occur. It suspends any organelles and contributes to the overall structure of the cell.

Ribosomes

Ribosomes are essential structures for all cells. They read discrete copies of information copied from DNA (discussed next) and use them to build a pro-tein by bonding a specific sequence of amino acids together. In doing so, the ribosomes participate in the anabolism of the cell and carry out dehydration synthesis reactions.

Deoxyribonucleic Acid

DNA is an essential macromolecule in every cell. It provides the necessary instructions for building all proteins required by the cell or the larger organ-ism to which the cell belongs. Recall that DNA is a nucleic acid made up of long stretches of nucleotides bonded together. It is characterized by the fact that it contains two chains of nucleotides that bond to each other and coil into a double helix.

DNA is read in discrete regions called genes; each gene is copied to trans-fer the instructions for building a given protein. First, the instructions are put into an RNA copy, and then the RNA molecule transfers the instructions to the ribosome for actual protein synthesis.

Much of life on Earth comprises prokaryotic cells, namely all bacteria. Because they are all unicellular and too small to be seen with the naked eye, they are often forgotten when we consider biodiversity. Due to their exces-sively small size (bacteria are, at most, one 10th the size of a eukaryotic cell), they are actually the most abundant life-forms on our planet. We don't gen-erally like to acknowledge the fact that the bacteria living on us and in us are greater in number than our own body cells. As eukaryotes, we all originated from prokaryotes, a fact that will be discussed in detail in Chapter 9.

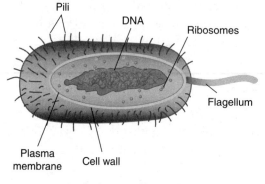

Prokaryotic cell structure

 Students often assume that prokaryotes don't have any internal organelles or any membranes. It is important to remember that all cells must have a plasma membrane to provide a protective, regulatory barrier between the external environment and the internal cytoplasm. Also, it is crucial to remember that all cells must have ribosomes so they can effectively decode the instructions held within DNA and put them to use building proteins. Most prokaryotes also have at least one protective cell wall, and some have short, protein-based pili that project from the cell wall and help with attachment to a surface or exchange of genetic information with other prokaryotes.

 ## Exercise 3.2

If you can successfully work through the following questions, you should feel confident about being able to go on to the next section.

1. The function of the ribosomes within a prokaryote is to _____.
    a. conduct homeostasis
    b. synthesize proteins
    c. store wastes
    d. regulate growth

2. Given that all prokaryotic organisms are unicellular, which of the following could a prokaryote never possess?
    a. An organic macromolecule
    b. An organelle
    c. A CHON atom
    d. An organ

3. Use the theme of continuity and variation to describe the concept of DNA and ribosome function within a prokaryotic cell.

_____

_____

_____

The Eukaryotic Cell

You will recall that a eukaryotic cell possesses more structural complexity than a prokaryote due to its internal, membrane-bound organelles that can specialize in structure and function. While many unicellular organisms are prokaryotic (such as all bacteria), others are eukaryotic, as in the case of most protists and some fungi. Unicellular eukaryotes must maintain a generalist approach, meaning that the single cell must be able to fully meet all the characteristics of life. Internally, however, unicellular eukaryotes are more specialized than bacteria, for they have a multitude of organelles, each highly efficient at a given process.

Nucleus

The large, easily visible nucleus makes eukaryotic cells readily identifiable from their prokaryotic counterparts. The nucleus is a complex structure bound by a nuclear envelope composed of a double membrane. Along the length of the envelope are nuclear pores, openings to allow for movement of molecules and thus communication between the inside of the nucleus and the cell's cytoplasm. Just inside the envelope, *nucleoplasm* suspends the DNA. This molecule stores and protects the genetic code, providing blueprints for the construction of many different proteins that constitute the organism's traits. The region where the DNA is tightly condensed and in the process of building *ribosomal RNA* (rRNA), which physically makes up the majority of a ribosome, is known as a *nucleolus*.

Endoplasmic Reticulum

The *endoplasmic reticulum* (ER) consists of a series of folded membranes with internal channels. In some places, the ER includes many embedded ribosomes within the membranes; this is called *rough ER*. In other places, the ER is free of any ribosomes, so it is called *smooth ER*.

Within the series of channels along rough ER, proteins synthesized at a ribosome are then packaged into a membrane-bound *vesicle* as a piece of the ER's own membrane is re-formed around the proteins. The vesicle is tagged and shuttled to the Golgi body.

Different enzymes are stored within the smooth ER. These enzymes help in many metabolic processes, including the synthesis of different lipids and carbohydrates.

Golgi Body

The *Golgi body* (or Golgi apparatus) is, like the ER, a series of flattened, membranous sacs stacked upon one another. The membrane of an incoming vesicle from the rough ER can fuse with the membrane of the Golgi, allowing transport of the protein into the sac of the Golgi. Within the boundaries of the Golgi, the protein is sorted, modified, and packaged either for release from the cell or for local use. The Golgi also helps to produce small organelles called *lysosomes* and to transport certain lipids around the cell.

Mitochondrion

The *mitochondrion* is often referred to as the powerhouse of the cell because of its crucial role in cell metabolism. It is bound by a set of double membranes, and the internal membrane actually extends out into long folds called *cristae*. These act to increase the surface area of the inner membrane, the primary site of energy transfer within the organelle. Inside the mitochondrion is an aqueous, cytoplasm-like solution called *stroma*, in which other metabolic reactions occur facilitated by certain enzymes. The overall function of the mitochondrion is *ATP synthesis*, as it captures energy from the bonds in glucose to catalyze the synthesis of ATP from its monomer subparts (adenosine diphosphate, or ADP, and inorganic phosphate, or P_i).

Unique Properties of Plant and Animal Cells

Plant cells can easily be distinguished from animal cells because they possess a few special structures. The *cell wall* is a rigid structure external to the cell membrane of a plant cell and composed of a polysaccharide called *cellulose* (which we know as dietary fiber, because it is indigestible). Inside the cell, two organelles also set the plant cell apart from its animal counterpart. The large *central vacuole* is a membrane-bound sac that can often comprise the majority of the volume of a typical plant cell. Inside, it stores water;

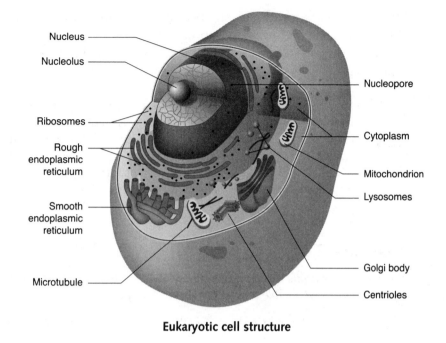

Eukaryotic cell structure

isolates some wastes; and contributes to the maintenance of *turgor pressure*, the force exerted from the cytoplasm against the cell wall. Finally, the *chloroplast* is an organelle lined with two membranes and containing a high concentration of pigments. The pigments are essential in trapping light to provide the energy for photosynthesis (described in detail in Chapter 4).

Other organelles are also common in eukaryotic cells. These include small, membranous containers called *lysosomes* that house digestive enzymes for intracellular digestion. Similar organelles that contain distinct enzymes are *peroxisomes*, named for the hydrogen peroxide they produce when breaking down alcohols. They also help to reduce free radicals that can be destructive to molecules.

Some structures are important in maintaining the internal framework and shape of the cell; these contribute to the eukaryotic cytoskeleton. The *cytoskeleton* is composed of different types of long chains of proteins that make up one of two basic fibers—a hollow microtubule or a solid microfilament.

Finally, other structures are present only in animal cells when they are undergoing the process of cell division. These structures are called *centrioles* and are actually composed of some of the same proteins that make up the microtubules just discussed. Two short microtubules are arranged at a 90-degree angle and move to opposite ends of a cell to create poles. They provide an anchor toward which a set of duplicated chromosomes is pulled before one cell splits in two.

Often, animal cells are overcharacterized as the cell type that contains mitochondria, implying that plant cells do not. This is not the case, however; plant cells need mitochondria to convert the glucose they produce during photosynthesis to ATP so as to power all of their life processes! Plant cells do typically possess fewer mitochondria than comparable animal cells, for they are less active and therefore have lower metabolic needs.

Exercise 3.3

If you can successfully work through the following questions, you should feel confident about being able to go on to the next section.

1. A eukaryotic cell can be distinguished from a prokaryotic one by the presence of _____.
   a. a membrane
   b. DNA
   c. internal membranes
   d. many ribosomes

2. If centrioles are observed in a given cell, then that cell is _____; if a central vacuole is observed instead, the cell might be described as

   _____.
   a. plant; animal
   b. animal; bacterial
   c. bacterial; plant
   d. animal; plant

3. Use the theme of the relationship between structure and function to describe the concept of the endomembrane system, which includes the nuclear envelope, the ER, the Golgi body, and the cell membrane.

   _____

   _____

   _____

4

Cell Homeostasis and Metabolism

☑ Homeostasis and metabolism are two essential characteristics of life; thus, each must ultimately occur on the cellular level.

☑ *Cell homeostasis* is carried out primarily through the plasma membrane and its intricate phospholipid bilayer structure. This allows a cell to be selectively permeable and to control movement into and out of its boundaries.

☑ Some movement across a membrane is passive and requires no energy investment by the cell. Other movement is considered active and does require an energy input in the form of ATP.

☑ *Cell metabolism* is essential because a cell has a constant supply of energy and an ability to use it for different functions.

☑ Plant cells are characterized by chloroplasts to enable *photosynthesis*, a pathway characteristic of *photoautotrophs* that can use the sun's energy to build organic food.

☑ Animal cells are characterized by the presence of relatively more mitochondria than plant cells. These organelles enable the cells to break down glucose in a controlled fashion so that much of the cells' internal energy can be extracted and stored temporarily in the form of ATP until it is ready to be used.

Cell Homeostasis

The chemical composition of the plasma membrane determines which substances are permitted to cross its boundary and either enter or leave a given cell. All membranes are mainly composed of a particular type of lipid molecule known as the *phospholipid*; these phospholipids are primarily responsible for regulating transport of substances across the membrane. Other protein molecules are also involved, allowing specific substances to pass that would otherwise be obstructed by the phospholipids.

Phospholipids

Each phospholipid is composed of a hydrophilic head and two long hydrophobic tails. Each head is made up of a three-carbon, glycerol-based backbone with a phosphate group attached to one of the terminal carbons. The negatively charged phosphate group provides the basis for the head's hydrophilic nature. Each of the two tails is made up of a long, nonpolar hydrocarbon chain; collectively, the tails are very hydrophobic.

Phospholipid Bilayer

The phospholipid bilayer structure of the plasma membrane allows the membrane to act as a physical and chemical filter. The heads of the phospholipids are tightly packed within each layer, so they are well suited for mechanically screening out anything larger than a small molecule.

Also within each layer, the heads are oriented toward either the outside or the inside of the cell, leaving the tails of the phospholipids all pointing toward the interior of the cell membrane. In other words, the cell membrane is structured somewhat like a sandwich: hydrophilic heads on the outside (chemically ideal for interacting with an aqueous environment); two layers of hydrophobic tails (perfect for discouraging passage of ionic or polar substances); and another layer of hydrophilic heads (ideal for interacting with the internal cytoplasm of the cell).

This selective permeability of the plasma membrane means that only small, nonpolar molecules such as oxygen (O_2) and carbon dioxide (CO_2) can rapidly and readily diffuse across it. Although polar, water is small and concentrated enough to diffuse through the bilayer as well.

Membrane Proteins

Proteins are typically scattered throughout the phospholipid bilayer. Some are called *peripheral*, because they are only partially embedded in the bilayer and are exposed only to either the external environment of the cell or the internal cytoplasm.

Other proteins, the *integral* or *transmembrane proteins*, are fully embedded within the bilayer and have access to both the external environment and the internal cytoplasm. These proteins can thus act as a channel and be useful in the movement of substances into and out of the cell that cannot diffuse through the phospholipid bilayer.

The specific types and quantities of proteins dispersed depend heavily on the particular functions of the cell type. For example, a cell with high metabolic needs may have a high concentration of integral proteins that recognize glucose and transport it into the cell.

Plasma Membrane

The collective structure produced by the phospholipid bilayer and the dispersed membrane proteins is called the *plasma membrane* (or cell membrane). This membrane can be the most external structure of a cell; it is therefore crucial in protecting the cell and in regulating the transport of substances across its surface.

Some additional structures are also typically scattered throughout the eukaryotic plasma membrane, most notably molecules of *cholesterol*. A steroid-based lipid, cholesterol associates with the hydrophobic tails and contributes to membrane rigidity. It also helps the membrane resist freezing (and potentially fracturing) at very low temperatures.

It should now make sense how and why the plasma membrane is essential to cell homeostasis. Its primary internal structure—the phospholipid bilayer—acts to keep most molecules on one side of the membrane. The integral transport proteins are specific to given substances; many are gated and thus open only when they receive a specific environmental signal. All of this contributes to the regulation of the cell's internal cytoplasm and works to keep the internal organelles in homeostasis and functioning optimally.

If the cell is part of a multicellular organism, then cell homeostasis might be important for the function of the tissue of which it is a part. If many cells are out of homeostasis, then the tissue may be dysfunctional, and the entire organ may ultimately be affected. This, in turn, may contribute to a disease or disorder for the individual organism.

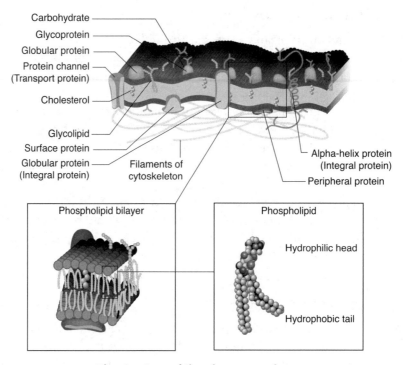

Carbohydrate
Glycoprotein
Globular protein
Protein channel
(Transport protein)
Cholesterol

Glycolipid
Surface protein
Globular protein
(Integral protein)

Filaments of
cytoskeleton

Alpha-helix protein
(Integral protein)
Peripheral protein

Phospholipid bilayer

Phospholipid

Hydrophilic head

Hydrophobic tail

The structure of the plasma membrane

Active and Passive Transport

Cells have two main ways of moving substances across the membrane: passive transport and active transport. *Passive transport* does not require an ATP investment by the cell, but *active transport* does. Passive transport includes the process of simple diffusion, which relies on the movement of substances down their concentration gradient (that is, from high concentration—where they readily bump into each other and transfer kinetic energy—to low concentration). Simple diffusion takes place right across the phospholipid bilayer as small, typically nonpolar substances squeeze through the phospholipids. A specialized form of simple diffusion is called *osmosis*, which involves the diffusion of water.

Other diffusion requires the assistance of integral proteins embedded in the bilayer. These proteins shield moving substances from the phospholipids and thus allow transport of larger, polar, or ionic substances. This type of diffusion is called *facilitated diffusion*. Since it still moves substances down

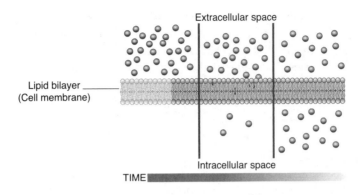

Simple diffusion across a membrane

their concentration gradient, it is still classified as passive transport. A specialized form of facilitated diffusion occurs across ion channels, which are transmembrane proteins that recognize a specific ion and transport it across the membrane.

Active transport requires ATP investment by the cell doing the transporting. This transport has two main types: cell membrane pumps and bulk flow. *Cell membrane pumps* require the assistance of an integral protein, but unlike facilitated diffusion, the protein requires ATP to change its shape and allow the movement of specific substances across. The most typical cell membrane pump in animals is the sodium-potassium pump, as this lines nerve cell membranes and is essential to their ability to maintain cellular homeostasis. The sodium-potassium pump moves three sodium ions (Na^+) from the cytoplasm to the external environment and then moves two potassium ions (K^+) from outside the cell to the internal cytoplasm.

Bulk flow, another type of active transport requiring ATP, involves the movement of many molecules into (*endocytosis*) or out of (*exocytosis*) a cell. If a cell is trying to ingest many solid particles or an entire cell in a form of endocytosis known as *phagocytosis*, it moves its membrane around the substance to create a pouch. The membrane eventually meets at the other side and pinches off completely to form a vesicle—a small, membrane-bound container now within the cell's cytoplasm. A similar process called *pinocytosis* occurs when a cell ingests a large volume of fluid and dissolved solutes.

Students often confuse the plasma membrane with the phospholipid bilayer. It is important to remember that although the plasma membrane consists mainly of phospholipids, it also contains many peripheral and integral proteins. This means that a substance might be able to diffuse across the plasma membrane but cannot diffuse across the selectively permeable phospholipid bilayer because it uses an integral protein to shield itself from the bilayer.

 ## Exercise 4.1

If you can successfully work through the following questions, you should feel confident about being able to move on to the next section.

1. Which of the following statements regarding the structure and function of the plasma membrane is true?
   a. The head of each phospholipid is hydrophobic.
   b. The heads of the phospholipids point outward toward the external environment.
   c. The tails of the phospholipids are oriented to face the internal cytoplasm.
   d. The tail of each phospholipid is hydrophilic.

2. Which of the following molecules would likely be too large and/or polar to diffuse through the bilayer and would instead require the assistance of a protein?
   a. Carbon dioxide
   b. Water
   c. Glucose
   d. Oxygen

3. Use the theme of regulation: feedback and response to describe the concept of the cell membrane and its role in cellular homeostasis.

   _____

   _____

   _____

Photosynthetic Metabolism

One of the most important chemical processes on Earth is that of photosynthesis. Carried out exclusively by autotrophs, photosynthesis allows for the production of carbohydrates, an essential energy source for all cells, using only the powerful and ongoing radiant energy emitted from the sun.

$6H_2O + 6CO_2 + Light \rightarrow C_6H_{12}O_6 + 6O_2$

This is the sum reaction for the process of photosynthesis. That is, a plant needs six molecules of water and six molecules of carbon dioxide, along with the energy trapped in its pigments, to build glucose and produce the oxygen gas by-product. This reaction does not in any way describe how photosynthesis occurs.

Light Reaction Step 1: The Electron Transport Chain

If light strikes the chloroplast, it is absorbed by the pigment molecules that are highly concentrated within structures called *photosystems*. These photosystems are found throughout the internal membranes of the chloroplast, called *thylakoids*. When the energy from light is absorbed by a specific photosystem called *photosystem II* (PS II), electrons within the pigments become excited (have high energy) and begin to move. They are picked up by the first in a series of proteins also embedded in the thylakoid membrane called the *electron transport chain* (ETC).

The first electron carrier of the ETC accepts the electron by participating in an oxidation-reduction, or *redox*, reaction. The pigment molecule within the photosystem gives up the excited electron and becomes oxidized. In turn, the electron carrier molecule that accepts the electron becomes reduced. This is actually just the first in a series of redox reactions, each moving the electron a little farther down the chain. Eventually, the electron reaches another photosystem—*photosystem I* (PS I).

Here, light of a slightly different wavelength is absorbed again by highly concentrated pigment molecules. The electrons moving in from PS II thus replace the newly excited electrons in PS I. These excited electrons move down a second electron transport chain, ultimately reaching the final electron acceptor. This molecule is ultimately embedded in the membrane to hand off the electron to another reaction waiting to take place just outside the thylakoid (that is, within the stroma of the chloroplast).

In the aqueous stroma, a molecule called $NADP^+$ is waiting near the final electron acceptor. It grabs the electron, along with an available hydrogen ion (H^+) in the stroma, to produce the molecule called NADPH. This molecule is an energy intermediate and will eventually be used to build a molecule of glucose in the dark reaction phase.

Light Reaction Step 2: Splitting Water

Within the chloroplast, as just discussed, short and repetitive extensions of the inner membrane create stacks of membranous sacs called thylakoids. Within each thylakoid, enzymes that catalyze the splitting of water are present in high quantities. There, two water molecules are broken apart at one time and converted into three products: the two oxygen atoms are joined together to produce one molecule of O_2; the electrons are stripped away from the four remaining hydrogen atoms, producing four hydrogen ions ($4H^+$) and four free electrons ($4e^-$). To summarize:

$$2H_2O \rightarrow 4H^+ + 4e^- + O_2$$

Light Reaction Step 3: Chemiosmosis (or Photophosphorylation)

The products of the water-splitting reactions are in turn used for the last step of the light reactions. The H^+ starts to build up within the thylakoids. It cannot diffuse across the phospholipid bilayer, but it can move through a H^+ ion channel. Luckily, this H^+ ion channel has a secondary (but very important) function: it is also an enzyme that catalyzes the synthesis of ATP from its monomer subparts, ADP and P_i. This essential molecule is named for its enzymatic function—*ATP synthase.*

As the H^+ ion passively diffuses through ATP synthase (and thus out of the thylakoid and into the stroma), it also kick-starts the catalysis of ATP that is waiting to take place. This collective process is known as *chemiosmosis,* the production of ATP using energy gathered from the passive transport of H^+ ions down their gradient.

Chemiosmosis works because the substance moving through the enzyme is an ion, and its positive charge causes the ATP synthase to experience a conformational (shape) change. This change in turn moves ADP and P_i into the optimal energetic positions, and they react to produce ATP. The ATP produced is ultimately used to make glucose in the next phase of photosynthesis, the dark reactions. The H^+ ion that diffused through ATP synthase also has another function: once in the stroma, it can be used to build NADPH at the end of the ETC (from step 1).

Dark Reactions: The Calvin Cycle

The Calvin cycle consists of a series of reactions; it starts with the incorporation of inorganic CO_2 and ends with both the desired product of glucose and the regeneration of the molecule needed to start the next round of the Calvin cycle (by combining with more CO_2).

Three molecules of CO_2 are used for each round of the Calvin cycle. Each CO_2 combines with a recycled organic molecule called RuBP. By bonding with five-carbon RuBP, the CO_2 becomes part of a six-carbon organic molecule and is thus "fixed." The total products are therefore three six-carbon molecules.

The three six-carbon molecules immediately split in half, producing six three-carbon molecules. From there, another reaction occurs to convert all of the new molecules into a different one, namely G3P. (G3P is also a three-carbon molecule, but it can be approximated as half of a glucose molecule.) To power the reaction in the Calvin cycle that finally makes G3P, both ATP and NADPH must be invested. Finally, all the actions of the light reactions are put to work!

This step creates enough G3P to do two things. Of chief importance, one G3P molecule is removed from the cycle per turn and will be used to build a glucose molecule. Also, five G3P molecules remain in the cycle and, with the investment of more ATP from the light reactions, are ultimately converted into three molecules of RuBP. The Calvin cycle is now ready to begin again.

With one more round of the Calvin cycle, starting with a three-CO_2 investment and ending with the removal of a G3P molecule, one glucose molecule can finally be produced.

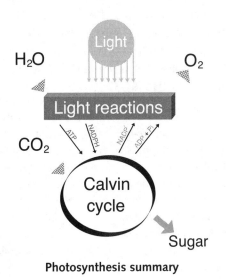

Photosynthesis summary

Photosynthesis is arguably the most significant metabolic pathway on Earth. Its glucose product makes up at least 70 percent of the total dry mass of a typical plant and thus acts as the basis of the food chain in most ecosystems. Additionally, the O_2 by-product of photosynthesis is essential for most organisms to metabolize efficiently. Nevertheless, plants are not responsible for the majority of the oxygen found on this planet; the nod goes instead to protists like algae and phytoplankton. These protists collectively produce approximately 70–80 percent of the Earth's atmospheric oxygen.

All *photoautotrophs*, organisms that produce their own food using energy from light, carry out the light reaction phase of photosynthesis as already described. Some plants, however, have evolved alternatives to the dark reactions in response to particular environmental conditions. Some plants, like corn and crabgrass, live in warm, arid conditions and carry out an alternative C_4 pathway in a water-conserving mechanism. Called C_4 *plants*, they can close many of their *stomata*, tiny pores clustered on the underside of leaves, when they need to stop water from escaping. The low CO_2 level that results activates an enzyme that incorporates any CO_2 into a four-carbon molecule, maintaining a strong gradient. This four-carbon molecule is transported to other cells and then broken back down into CO_2 so the Calvin cycle can take place and glucose can still be built. (Note that more typical plants are called C_3 *plants* because of the three-carbon intermediate that is utilized to build glucose in the Calvin cycle.)

Some plants, like cacti and succulents that have evolved in more extreme desert ecosystems, use a different mechanism still. They need to close all of their stomata during the day to conserve water, but this in turn means that no CO_2 can diffuse in during the day when the light reactions are ready to work. These plants, called *CAM plants*, instead incorporate CO_2 into a variety of organic molecules at night when the stomata are open and then release CO_2 from those molecules as needed during the day when the light reactions are producing NADPH and ATP.

 It is easy to confuse CAM metabolism with C_4 metabolism given that both evolved as water-saving mechanisms. If you can remember that CAM plants are desert plants that live in extremely arid environments, then you will know that they have to take more extreme measures than even C_4 plants. CAM plants behave in the opposite manner to C_3 and C_4 plants in that they close their stomata completely during the day and open them at night. This is only because they have evolved a special means to store CO_2 overnight when they don't also risk losing too much water. They then release CO_2 from storage so it can be used in the Calvin cycle to build glucose during the day when sunlight is abundant.

Exercise 4.2

If you can successfully work through the following questions, you should feel confident about being able to move on to the next section.

1. Which of the following are products of the light reactions and reactants of the dark reactions of photosynthesis?
   a. NADPH and glucose
   b. NADPH and ATP
   c. ATP and glucose
   d. CO_2 and glucose

2. Which of the following is the actual desired product of the photosynthetic pathway, and which is considered a by-product?
   a. Oxygen gas; water
   b. Water; glucose
   c. Glucose; water
   d. Glucose; oxygen gas

3. Use the theme of energy transfer to describe the concept of the light and dark reactions of photosynthesis.

   _____

   _____

   _____

Cellular Respiration Metabolism

Along with photosynthesis, the process of cellular respiration is one of the most essential on Earth. While photosynthesis is responsible for the production of glucose, cellular respiration describes a way to slowly and efficiently break down that glucose to release ATP molecules for use within the cell.

$C_6H_{12}O_6 + 6O_2 \rightarrow 6H_2O + 6CO_2 + ATP$

This is the sum reaction for the process of cellular respiration. That is, a cell needs six molecules of O_2 and one molecule of glucose to produce the desired energy product, 36 to 38 molecules of ATP. Six molecules of water and six of CO_2 are also released as by-products. Just as with photosynthesis, the sum reaction does not in any way describe how cellular respiration actually occurs.

Cell Respiration Step 1: Glycolysis

Glycolysis involves the slow breakdown of the glucose molecule and takes place right in the cytoplasm of the cell with the coordinated help of several enzymes. Glucose is a stable molecule, so it needs to be made unstable before it can be broken down. Two ATP molecules are actually invested here, each one contributing a P_i to the glucose. This energy transfer makes the glucose molecule unstable so it is ready to break down and release even more energy.

As glucose is partially broken down, most of the released energy is stored in the form of NADH. (This energy intermediate is analogous to the NADPH produced in the light reactions of photosynthesis.) This occurs through another redox reaction; a NAD^+ molecule grabs an electron from the organic molecule along with a H^+ ion to produce NADH. Any NADH molecules produced will eventually be converted into ATP energy during the final phase of cellular respiration.

The rest of the energy captured from the breakdown of glucose is recaptured directly in the form of ATP. Four total ATP molecules are produced during glycolysis, so, considering the two ATPs that were invested at the start, the process results in a net gain of two ATP molecules.

The final chemical products of glycolysis are two molecules of three-carbon pyruvic acid. Much of the energy in glucose has been released at this point (some has been captured by the cell in the form of ATP and NADH), but some still remains in the pyruvic acid.

Cell Respiration Step 2: The Citric Acid Cycle (Krebs Cycle)

The three-carbon pyruvic acid product of glycolysis is transported into the mitochondrion, specifically into the inner mitochondrial matrix. Here, it participates in another reaction during which it is broken down into two-carbon acetyl coenzyme A (CoA), releasing the other carbon present in pyruvic acid in the form of CO_2. This step also results in the formation of an additional molecule of NADH.

The initial glucose molecule produced two pyruvic acid molecules, so two pyruvic acid molecules produce two molecules of acetyl CoA, two molecules of CO_2, and two molecules of NADH.

Now each acetyl CoA molecule enters the Krebs cycle, which also takes place in the mitochondrial matrix. Here, the two-carbon acetyl CoA combines with a recycled, four-carbon organic molecule called OAA to produce six-carbon citric acid. This reaction signals the start of a series of reactions forming the cycle, for citric acid is immediately converted into yet other intermediates until OAA is finally regenerated, and another round of the cycle can continue.

What is the point of all these chemical reactions? Along the way, many of them also generate additional NADH molecules. Others, in the same spirit, generate another form of an energy intermediate known as $FADH_2$. It is collected, as is NADH, until it can be used to produce ATP in the final step. Some ATP is directly generated as well, specifically one molecule of ATP per one molecule of acetyl CoA (or two ATPs per glucose).

Cell Respiration Step 3: The ETC and Chemiosmosis

All of the NADH produced in glycolysis, the conversion of pyruvic acid into acetyl CoA, and the Krebs cycle is now ready in the matrix. The same is true for the $FADH_2$ produced solely in the Krebs cycle. Both are accessible by proteins embedded in the inner mitochondrial membrane that make up another ETC. The primary electron acceptor in this ETC grabs an electron from the H atom in a molecule of NADH and regenerates the NAD^+ needed for more glycolysis to occur. The H^+ ion that results is actively pumped across the inner mitochondrial membrane and into the intermembrane space. This is done to create a gradient of H^+ ions across the membrane (which should sound familiar!).

The electron accepted by the chain moves down the ETC, reacting again in a series of redox reactions, until it reaches the final electron acceptor. That molecule is O_2 itself. As one molecule of O_2 accepts four electrons, it also combines with four available H^+ ions to produce two molecules of water. (This is the reverse reaction of water splitting during photosynthesis.)

Cell respiration is another essential metabolic pathway. Most organisms on Earth are aerobic, using O_2 to efficiently break down glucose and generate ATP. But how do anaerobic organisms go about producing ATP? And what happens when O_2 levels are low for aerobic organisms? These situations require a secondary pathway, namely that of *fermentation*. Fermentation, just like cellular respiration, begins with glycolysis. In fact, the only ATP molecules generated from fermentation are made during the glycolysis step. After the two molecules of pyruvic acid are created at the end of the pathway, they must be converted into another, less toxic substance. Depending on the organism, one of many pathways can take place. The two most common involve the final product of either lactic acid or ethyl alcohol.

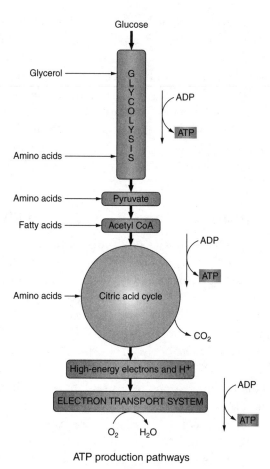

ATP production pathways

Cellular respiration summary

It is helpful to review the net gain of ATP during each step of cellular respiration and fermentation. Remember that both processes begin with glycolysis and the slow breakdown of glucose into two pyruvic acid molecules. Glycolysis results in a net gain of two ATP.

Cellular respiration continues by first converting the pyruvic acid into acetyl CoA, a step not directly associated with ATP gain, and then the acetyl CoA enters the Krebs cycle. There, each acetyl CoA is linked to the production of two ATPs, meaning that four total ATPs per glucose molecule result from the Krebs cycle. Finally, the action of the ETC and, more directly, that of chemiosmosis are responsible for the production of 30 to 32 ATPs per glucose. Cellular respiration accounts for a total of 36 to 38 ATPs per glucose molecule initially invested.

If a cell is anaerobic and/or doesn't contain mitochondria, then it is restricted to fermentation. This is less efficient, because the cell only gains the two ATPs that result from glycolysis. No additional ATP comes from the conversion of pyruvic acid into either ethyl alcohol (in alcoholic fermentation) or lactic acid (in lactic acid metabolism).

It is easy to confuse the NADPH used in the photosynthetic pathway with the NAD used in the cellular respiration pathway. Remember that the energy shuttle with a *P* in its name is related to the pathway that starts with the same letter—photosynthesis. By default, that leaves NAD to be associated with cellular respiration.

Exercise 4.3

If you can successfully work through the following questions, you should feel confident about being able to move on to the next section.

1. The vast majority of the ATP gained during cellular respiration is directly produced during which of the following steps?
   a. Glycolysis
   b. The Krebs cycle
   c. The ETC
   d. Chemiosmosis

2. Which of the following *incorrectly* pairs the metabolic process with its specific location within a cell?
   a. Glycolysis—cytoplasm
   b. Krebs cycle—mitochondrial matrix
   c. Fermentation—mitochondrial matrix
   d. ETC—inner mitochondrial membrane

3. Use the theme of continuity and variation to describe the concept of the breakdown of glucose during glycolysis and either fermentation or cellular respiration.

_____

_____

_____

5

The Cell Cycle
and Cell Division

☑ Given that an individual cell may itself constitute an organism, it must be able to carry out the process of *cell division* to enable reproduction, one of the six characteristics of life. When cells within a multicellular organism divide, they are not contributing to reproduction but instead are contributing to the growth of the organism.

☑ The life span of a cell, from the point it is created from cell division to the time it finally divides to create new offspring cells, is called the *cell cycle*.

☑ The division of the DNA within the nucleus of eukaryotic cells can be complicated. The process of *mitosis* carefully coordinates the process, ensuring that exact copies of the genome reach each offspring cell produced.

The Cell Cycle

The events of the cell cycle represent specific, coordinated periods of growth, replication of DNA, and eventual division of the cell. Checkpoints along the way help to ensure that the entire process is carefully regulated and that a specific step takes place only when the cell is well prepared.

Interphase: G_1 Phase

When a new cell has been created, it begins its life in the *Gap 1* (G_1) *phase* of the cell cycle. The term *gap* is something of a misnomer, as many activities happen during this phase. The cell metabolizes, grows, maintains homeostasis—basically carries out life's daily functions. The name is actually a reference to the fact that nothing directly related to cell division occurs during G_1. However, some energy that will be helpful in cell division is gained.

The G_1 phase is the first of three sequential phases that together compose a larger one called *interphase*. The name literally means "time between divisions," so *interphase* describes any part of the cell's life cycle when it is not dividing. During this time, the DNA in the nucleus is found in the disorganized and less visible chromatin form. It only coils up into the more recognizable chromosomes for mitosis.

Interphase: S Phase

After the cell has sufficiently increased in size and stored some ATP energy, it enters the *synthesis* (S) *phase*. During the S phase, the DNA in the nucleus of the cell is copied perfectly in a process called *DNA replication*; the DNA must remain in chromatin form for its nucleotides to be read easily so this process can occur. The DNA is copied in preparation for an upcoming division; the newly created cell will need an exact set of genetic instructions from the parent cell to provide continuity to life between generations.

Actually, before a cell enters the S phase, it must pass a checkpoint; if the cell is not in appropriate condition and/or is in a state of stress, then it does not enter the S phase. It may instead enter a perpetual gap phase called G_0 (pronounced G naught). Here, the cell continues to carry out normal cell business but does nothing specific to prepare for division. If conditions become appropriate for the cell to divide, then the cell can be called out of G_0 and enter the S phase.

Given that the S phase immediately follows G_1 and is not directly part of mitosis, it is also considered part of interphase.

Interphase: G₂ Phase

After the lengthy and energy-requiring process of DNA replication has occurred, the cell enters a second gap phase (G_2). Again, the cell is busy carrying out normal cell business but doesn't do much to prepare for division until the end of the phase. Then there is another checkpoint to regulate the process. The cell should only leave G_2 when conditions are appropriate and the cell is fully prepared to divide.

The G_2 phase is considered the last part of interphase. The DNA again remains in chromatin form so that it can be read and protein synthesis can occur (a process described in detail in Chapter 7). Once the cell leaves G_2, it enters the nuclear division process proper.

Mitosis: M Phase

A cell that enters mitosis (*M phase*) is committed to the division of the nucleus. It has already grown to sufficient size, stored some energy, and copied its DNA. Now it has to carefully coordinate the splitting apart of the two copies in a well-orchestrated series of steps. These sequential steps are called prophase, metaphase, anaphase, and telophase. To keep track of the duplicated DNA (the product of the S phase), at the start of mitosis, the chromatin begins to condense into *chromosomes*—tightly packed, well-organized bundles of DNA that each contain a unique set of genes. At the end of mitosis, once the duplicated sets of chromosomes have been separated and allocated to two new nuclei, the chromosomes uncoil back into chromatin to signal the start of the next cell cycle.

Cytokinesis

Cytokinesis is technically not part of mitosis, as it does not directly entail the division of the nucleus. Instead, it involves the division of the cytoplasm into two separate cells and the formation of new cell membranes (and cell walls, if applicable). Cytokinesis does not always occur along with mitosis (that is, multinucleate cells can be created, as is typical in skeletal muscle cells), but when it does, it generally occurs simultaneously with telophase.

Cell Cycle Regulation

It is essential to keep the cell cycle properly regulated so that an optimal number of cells is maintained to ensure efficient functioning of the organ-

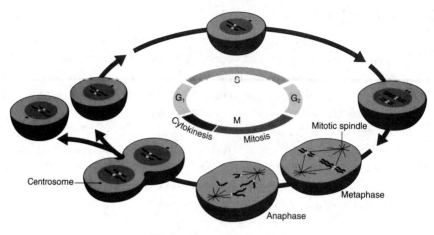

The eukaryotic cell cycle

ism. The G_1 checkpoint discussed previously partially contributes to the effort. It is considered a growth checkpoint, because if the cell is not large enough and doesn't have enough energy, the cell cycle will not continue into the next phase and enters the G_0 phase. Other checkpoints exist as well; these include one in the G_2 phase, during which DNA repair enzymes proofread the products of DNA replication in the S phase, and another toward the end of mitosis that signals the newly produced cells to complete the M phase and enter interphase and the next cell cycle.

When a cell loses control over the cell cycle, it is often due to a mutation of genes that produce the regulatory proteins operating at the checkpoints. When a cell loses all regulatory ability, *cancer* can result. Cancer, regardless of type, is always associated with an unintentional accumulation of cells that is eventually termed a *malignant tumor*. If all the cancerous cells can be killed or removed, then only healthy cells with appropriate regulatory ability should remain, and proper regulation of the cell cycle is restored. The person previously diagnosed with cancer is then said to be in remission.

 It is common to confuse the notion of cell division with the concept of mitosis. They are often used as synonyms when, in fact, they are not. Eukaryotic cell division includes the processes of both mitosis and cytokinesis. But mitosis can happen without cytokinesis to create specialized multinucleate cells.

If the cell cycle seems difficult to understand, remember that it is basically just a series of organized steps that allow a new cell to gain energy and nutrients; copy its DNA; gain more energy and nutrients (to recover from the replication and prepare for the upcoming step); and finally divide to create two new offspring cells.

Exercise 5.1

If you can successfully work through the following questions, you should feel confident about being able to move on to the next section.

1. Which of the following correctly links the phase of the eukaryotic cell cycle with its primary function?
   a. G_2—DNA replication
   b. G_0—nuclear division
   c. S—synthesis of proteins
   d. G_1—metabolism and growth

2. In a multicellular eukaryotic organism, cell division contributes directly to which of the following characteristics of life?
   a. Reproduction
   b. Growth
   c. Homeostasis
   d. Metabolism

3. Use the theme of regulation: feedback and response to describe the concept of the G_1 checkpoint in the cell cycle.

   _____

   _____

   _____

Mitosis: Nuclear Division

The M phase of the cell cycle is itself made up of a well-coordinated series of events. DNA must become highly organized and compact so that the DNA copies produced during the S phase can be split accordingly and appropriated to two separate nuclei.

Mitosis Step 1: Prophase

Prophase involves the cell preparing its internal structures for the upcoming nuclear division. First, the DNA begins to condense from the dispersed, threadlike chromatin into thick, ribbonlike chromosomes. Because of this, the region known as the nucleolus (where rRNA is synthesized) disassembles and is no longer visible.

At the same time, short, microtubular structures called centrioles assemble and begin to move toward opposite ends of the cell. Long microtubules also form and extend from each centriole toward the midline of the cell; some even extend from centriole to centriole. These spindle fibers act as the scaffolding that will eventually separate the two copies of DNA that resulted from the S phase. The nuclear envelope begins to break down as well to facilitate the movement of each of the copies of the genome to different regions within the cell.

Mitosis Step 2: Metaphase

In *metaphase*, the DNA is pulled toward the midline of the cell. Each chromosome, in duplicated form with sister *chromatids* (identical copies produced during the S phase) attached at a region called the *centromere*, is pulled by the spindle fibers until they line up single file along the midline of the cell. The midline, also called the *metaphase plate*, is established by the positioning of the centrioles during prophase. The centrioles move to opposite ends of the cell to establish two poles; the midline is thus positioned like an equator directly between. This means when the chromosomes are aligned on the metaphase plate, each sister chromatid is positioned so it faces one of the two centrioles.

Mitosis Step 3: Anaphase

While metaphase lined up every chromosome along the equator, *anaphase* actually does the work of pulling apart the sister chromatids. Given that each chromatid within a duplicated chromosome is already positioned facing one of the two poles, it is now pulled by the spindle fibers toward the centriole it is facing. This results in the sister chromatids being pulled in opposite directions after first being separated at the centromere. Each chromatid migrates toward each centriole.

Mitosis Step 4: Telophase

After the chromatids reach the poles, the nuclear division process is near completion. In *telophase*, the DNA—copies having been successfully separated—begins to uncoil back into chromatin form. At the same time, new nuclear envelopes begin to re-form around each copy of the cell's genome, and the centrioles and spindle fibers disassemble. If not only mitosis but full cell division is the desired outcome, then cytokinesis will be coordinated with telophase.

Cytokinesis

Recall that cytokinesis is not really part of mitosis but coincides with it to complete cell division. Cytokinesis differs markedly in plant cells and animal cells due to the more complicated cell wall structure in plants. In animal cells, the cell stretches during mitosis due to the extension of the spindle fibers. The membrane starts to pinch inward along the midline and creates a structure called a *cleavage furrow*. This is due to the contraction of concentric microfilaments within the membrane at the midline, similar to a drawstring hood cinching closed.

In a plant cell, the presence of a cell wall external to the membrane creates a more challenging structural situation. The cell must build a new cell wall from the inside out. It accomplishes this by using vesicles containing cellulose to deposit their contents in pieces along the midline, establishing the *cell plate*. This continues until the cell plate extends completely and fully separates the two new offspring cells. It then formally becomes a cell wall.

In eukaryotes, cell division involving mitosis takes place throughout the body of a multicellular organism to produce more body cells and contribute to growth. Body cells are more formally known as *somatic cells*; in humans, these include differentiated cells like skin cells and liver cells. Because somatic cells generate new somatic cells within an existing organism, and

| Prophase | Metaphase | Anaphase | Telophase | Cytokinesis |

Mitosis and cytokinesis overview

because sexually reproducing organisms receive one set of chromosomes from each parent during fertilization (a process that will be described in detail in Chapter 6), somatic cells within an individual are called *diploid* (having two sets of chromosomes for each chromosome type).

In prokaryotes, cell division is much simpler because there is no nucleus to contend with in accessing the DNA. In fact, the lack of a nucleus means that prokaryotes can't do mitosis at all. Instead, they carry out a process known as *binary fission* in which one parent prokaryote cell replicates its DNA (done quickly because prokaryotes only have a single, circular chromosome) and then splits up the replicated copy from its template and also divides the cytoplasm as it splits into two genetically identical offspring cells. Similar to what happens in eukaryotic cell division, the offspring cells each contain a copy of the original DNA and approximately half of the cytoplasm of the original parent cell.

It is very common to confuse the idea of a cell being characterized as diploid and a cell that has two copies of every chromosome. A cell that is diploid has two sets of chromosomes, one inherited from each parent. It is possible for a cell to have one copy of each chromosome and still be diploid, as is typical during G_1 of the cell cycle. After the cell passes the S phase, the DNA has been copied, and the cell then contains two copies of the diploid chromosomes until it divides during mitosis.

It is often helpful to summarize the overall process of mitosis so you can better understand the individual steps. Mitosis begins with one diploid parent cell containing a duplicated set of chromosomes. The cell goes through the orchestrated steps of prophase, metaphase, anaphase, and telophase to separate the two sets of chromosomes carefully and allocate them to opposite poles of the cell. At the end of mitosis, two separate nuclei have been created, one around each set of chromosomes. Because each of the chromosomes in the diploid set of DNA was carefully copied and the copies separated, the resultant offspring cells are also diploid. The main difference is that each offspring cell starts its cell cycle with a single copy of the diploid chromosomes.

Exercise 5.2

If you can successfully work through the following questions, you should feel confident about being able to move on to the next section.

1. During which of the following sequences are the events of interphase and mitosis correctly ordered?
   a. G_2 phase, S phase, prophase, telophase
   b. G_1 phase, S phase, metaphase, telophase
   c. S phase, G_2 phase, anaphase, metaphase
   d. G_1 phase, G_2 phase, metaphase, prophase

2. Which of the following is true regarding mitosis?
   a. It always is done as part of cell division.
   b. It results in two offspring cells, each with two copies of each chromosome.
   c. It results in two offspring cells, each with a diploid set of chromosomes.
   d. It occurs in eukaryotic cells and prokaryotic cells, although it is more complex in eukaryotes.

3. Use the theme of continuity and variation to describe the concept of the cell cycle as it relates to the different forms that DNA takes in interphase and mitosis.

   _____

   _____

   _____

6

Heredity: Meiosis and Mendelian Genetics

☑ Organisms that reproduce sexually have to divide their own genetic material in half before they can pass along their DNA to their offspring. This requires the production of specialized haploid cells in a process called *meiosis. Gametes*, or sex cells such as sperm and egg, must be haploid because they are designed to combine with another gamete in the fertilization process.

☑ Meiosis ensures that the number of chromosomes contained in a cell's nucleus is consistent from generation to generation and between different individuals that are part of the same species.

☑ The nuclear division process of meiosis orchestrates two divisions for the production of four offspring cells. It also introduces specific mechanisms that facilitate processes of genetic exchange.

☑ Although he knew nothing of meiosis or DNA at the time of his research, Gregor Mendel was able to understand the outcome of the genetic recombination taking place in meiosis and to conceive the mechanisms by which organisms inherit traits from their parents.

Meiosis: Production of Gametes

Meiosis occurs in a series of two sequential nuclear divisions, appropriately termed *meiosis I* and *meiosis II*. Meiosis I is more distinct from mitosis, because its mechanisms act to introduce genetic variation in an effort to increase genetic efficiency. This is counter to the mechanisms of mitosis, which act to preserve the genetic code by copying the entire genome precisely. Meiosis I also acts to reduce the chromosome number within the cell from diploid to *haploid* (having only one of each type of chromosome) by separating the sets of chromosomes, one received from each parent during fertilization. These pairs of chromosomes that exist in diploid cells are called *homologous*; they carry information for the same types of genes, but the specific form of the gene will vary if the parents have different resultant traits. Because meiosis I produces two haploid cells from one diploid parent cell, it is also called the *reduction division*.

Meiosis II, on the other hand, has many events in common with mitosis. For that reason, it is also called the *mitotic division of haploid cells*. The main difference between the two is that mitosis starts with one diploid cell, whereas meiosis II begins with two haploid parent cells. From there, the process of nuclear division within each haploid cell is basically the same as the process of nuclear division in mitosis. The chromosomes line up in single file along the midline; the sister chromatids are pulled apart and moved to the centrioles at opposite poles; and the new nuclei form. Because no type of mitotic division acts to reduce the chromosome count, the four cells that result from meiosis II remain haploid.

The first round of meiosis is the more important of the two as well as the one more distinct from mitosis. During meiosis I, homologous pairs of chromosomes are carefully aligned and an exchange of genes may occur. This genetic shuffling, along with the random alignment of the pairs along the equator that follows, is responsible for creating gametes that are genetically unique.

Meiosis Step 1: Prophase I

Prophase I is similar to prophase of mitosis in many ways that involve cell preparation for the upcoming division process. The DNA coils up and condenses into chromosomes (so any observed nucleoli are obstructed); the nuclear envelope breaks down; the centrioles form and migrate to opposite poles; and the parts of the spindle apparatus take shape. The key difference in prophase I is that, after chromosomes form, a process called *synapsis* occurs. During synapsis, homologous chromosomes pair up and form a *tetrad*. In this formation, crossing over may occur to exchange genes between *homologues*, or a homologous pair of chromosomes, thus increasing the potential for genetic diversity in the resultant gametes.

Meiosis Step 2: Metaphase I

Metaphase I involves the homologous pairs being pulled toward the metaphase plate by the spindle fibers and then aligned so that each pair straddles the equator. During this process, independent assortment occurs as one member of each homologous pair is randomly aligned on one side of the equator and thus randomly arranged with all the other chromosomes on that same side. As with crossing over, this serves to increase genetic diversity within the gametes that result, because all the chromosomes that happen to be positioned along one side of the midline will end up in an individual gamete together and will be separated from those on the other side.

Meiosis Step 3: Anaphase I

Anaphase I continues the nuclear division process by separating the chromosomes within a homologous pair. All of the chromosomes on one side of the midline are collectively pulled—again by the spindle fibers—toward one of the two centrioles. The other set of chromosomes is collectively pulled toward the other pole; thus, the homologous pairs undergo the process of separation.

Meiosis Step 4: Telophase I and Cytokinesis

During telophase I, chromosomes reach the centrioles at the poles. New nuclear envelopes form around each haploid set of DNA, and cytokinesis begins to place the nuclei into individual cells. This completes the steps of meiosis I and results in two genetically unique haploid cells, each with two copies of the genome (the cell went through an S phase before starting meiosis).

Meiosis II picks up there and, as previously described, takes each of the haploid cells produced from meiosis I through a second, mitosis-like nuclear division. Prophase II involves the duplication of remaining centrioles and their migration toward opposite poles, the formation of the spindle fibers, and the breakdown of the nuclear envelopes.

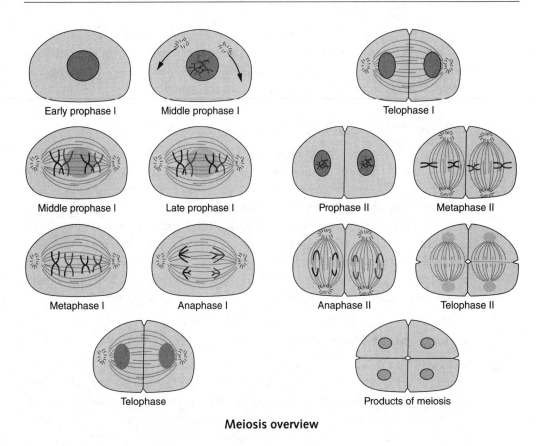

Early prophase I · Middle prophase I · Telophase I

Middle prophase I · Late prophase I · Prophase II · Metaphase II

Metaphase I · Anaphase I · Anaphase II · Telophase II

Telophase · Products of meiosis

Meiosis overview

Other Division Processes

While meiosis effectively accounts for the production of haploid cells, it does not completely account for the production of specialized gametes. *Spermatogenesis* describes the mechanism behind the creation of sperm cells (or spermatocytes), while *oogenesis* describes the production of egg cells (or ova).

In spermatogenesis, meiosis I and II proceed as expected. The cytokinetic events that take place at the end of each division split each parent cell roughly in half to create equal-sized offspring cells. The end result of spermatogenesis is then four haploid sperm cells of roughly the same size and composition. Each sperm also reduces its cytoplasm so that it barely surrounds the nucleus and develops a tail-like flagellum to allow it to propel itself through an aqueous environment in an attempt to reach the egg for fertilization.

Haploid (N)   Diploid (2N)

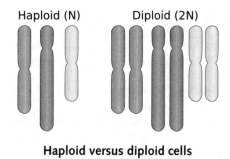

Haploid versus diploid cells

In oogenesis, the process is notably different. Primarily, the successive cytokinetic events at the end of each division are not carried out equally, meaning that one cell receives a majority of the cytoplasm and energy. At the end of meiosis, one large egg is created, and three small polar bodies result. While each of the four cells produced is haploid, only one is large enough to potentially support the growth of an embryo after fertilization. The other three haploid cells are called *polar bodies* and eventually break down.

 It is easy both to confuse and to misuse the terms *haploid* and *diploid*. Remember that they involve the number of unique sets of DNA an individual cell has. In other words, each term relates back to whether a cell had one or two parents and was a product of asexual or sexual reproduction.

 The need for a specialized nuclear division process like meiosis might seem difficult to understand. Remember that the main goal of meiosis is ultimately to create four genetically unique haploid cells. The cells produced must be haploid, for they are designed to be reproductive cells like gametes (or in some cases, other reproductive cells called *spores*). The entire purpose of a gamete like a sperm or an egg is to merge with a gamete of the opposite type in an attempt to produce a fertilized egg (or *zygote*). This zygote receives one set of chromosomes from each parent to become a new diploid cell, so the gametes that combine to create the zygote must themselves be haploid.

Exercise 6.1

If you can successfully work through the following questions, you should feel confident about being able to move on to the next section.

1. Which of the following *incorrectly* characterizes a specific step in meiosis?
   a. Prophase I—Crossing over occurs.
   b. Metaphase II—Independent assortment occurs.
   c. Telophase II—Four haploid nuclei are created.
   d. Telophase I—Two haploid nuclei are created.

2. Which of the following correctly characterizes a cell in metaphase I and/or metaphase II?
   a. A cell in metaphase I is haploid, while one in metaphase II is diploid.
   b. A cell in metaphase I has one copy of each set of chromosomes.
   c. A cell in metaphase II has two copies of each set of chromosomes.
   d. A cell in metaphase I is diploid, while one in metaphase II is haploid.

3. Use the theme of the relationship between structure and function to describe the difference between haploid and diploid cells.

   _____

   _____

   _____

Mendelian Genetics

Gregor Mendel is one of the most significant figures in the history of genetics and is aptly referred to as the father of modern genetics. Without any knowledge of DNA or the gene, Mendel uncovered the fundamental principles that govern heredity, the transmission of traits from parents to offspring. His efforts will be discussed in detail later in this chapter.

Trait

A *trait* is a characteristic that is genetically determined in an organism. In Mendel's experiments with pea plants, he observed traits such as flower color, flower position, stem height, and pod shape.

Allele

An *allele* is a particular form or expression of a gene. In his genetic crosses with pea plants, Mendel observed that the tall-stem-height allele in the presence of the short-stem-height allele within an individual organism would mask the latter when the organism expressed the tall-stem-height trait. He called the tall allele the *dominant* one and the masked allele the *recessive* one.

Genotype

A *genotype* is the combination of two alleles that any individual organism has for a given trait. As long as an organism is a product of sexual reproduction, it is expected to have two alleles for each trait. Having two of the same type of allele for a trait is described as *homozygous*. If both alleles are of the dominant type, then the genotypic designation is further specified as *homozygous dominant*; however, if both alleles are of the recessive type, then the genotype is called *homozygous recessive*. When the genotype contains a mix of one dominant and one recessive allele, it is described as *heterozygous*.

Phenotype

A *phenotype* is the physical expression of the genotype. As long as the genotype comprises at least one dominant allele for the trait (that is, is homozygous dominant or heterozygous), the phenotype expressed will correspond with the dominant allele and will also be described as dominant. Only when the genotype is made up of two recessive alleles will the organism show the recessive phenotype. Usually the genotype is a dependable predictor of the phenotype, but sometimes complex genetic patterns can emerge, or the environment can influence expression of a genotype and thus alter the phenotype.

Mendel's Research

Gregor Mendel was an Austrian monk who lived in the mid-1800s and spent much of his time tending to the monastery garden. He became fascinated by the transmission of traits from parents to offspring in the generations of pea plants he raised and observed that for most traits, two discrete forms of the trait were visible. He eventually developed a specific methodology, first to self-pollinate plants and establish pure strains and then to cross-pollinate plants with specific traits to carry out a huge variety of genetic crosses. Mendel later used the massive amount of data he collected over many generations for statistical analysis to help develop what would later be recognized as the laws of heredity. His contributions to biology remain so great that, to this day, he is referred to as the father of modern genetics.

Once Mendel had created a plant that was a pure strain for a specific trait of interest, that plant was then crossed with another plant that was a pure strain for the other form of the trait. For example, a pure-breeding, purple-flowered pea plant was cross-pollinated with a pure-breeding, white-flowered pea plant. He observed that the result of such a cross was always purple-flowered offspring plants; thus, the purple form of the trait was called dominant and the white form was recessive.

Especially fascinating to Mendel, however, was that the white-flower trait was not entirely lost from this resultant generation of purple-flowered plants. For example, when a purple-flowered plant from the first cross was cross-pollinated with another like plant, the result was three purple-flowered off-spring for every one white-flowered offspring produced. Mendel was onto something big!

The Three Laws of Heredity

Through Mendel's careful work and subsequent statistical analyses, he was able to uncover three laws of heredity:

1. The *law of dominance* states that if both a recessive allele and a dominant allele are present in the same genotype (that is, the genotype is heterozygous), then the dominant allele masks the expression of the recessive allele, and the phenotype of the organism is described as dominant.
2. The *law of segregation* describes the way in which only half of the genetic information possessed by either parent is passed along to any one offspring at a time. Although Mendel had no idea of DNA or meiosis, he was able to understand the outcome of the reduction division that takes place during meiosis I.
3. The *law of independent assortment* states that the specific inheritance of one trait has no effect on the inheritance of another. This is only true as long as the traits in question are housed on separate chromosomes. If they are in fact located on the same chromosome, the traits and genes are said to be linked, and the law of independent assortment does not hold true. Each of the traits that Mendel studied was actually stored in genes on separate chromosomes within the pea plant's genome, so he had no way of observing the effects of linked genes.

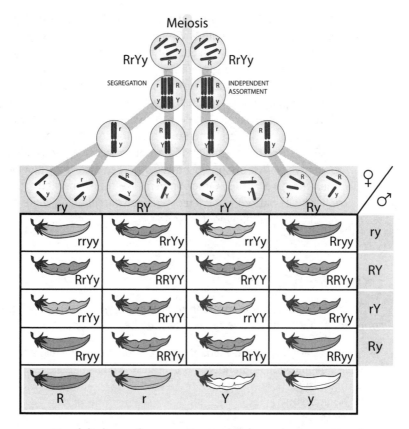

Mendel's laws of segregation and independent assortment

Mendel's laws of segregation and independent assortment can seem overwhelming. It is helpful to remember that the law of segregation describes an individual diploid cell having to separate the two alleles for a given gene before it creates a gamete to potentially pass the genetic information along. In other words, a diploid precursor cell must randomly split up its two sets of DNA (during meiosis) before it can produce an effective haploid gamete.

Independent assortment, on the other hand, has more to do with the fact that genes are not inherited on a gene-by-gene basis but rather are packaged together with a bunch of other genes on a chromosome. Inheritance really occurs on the chromosome level. Most genes are inherited independently from the rest, as long as they are not housed on the same chromosome (that is, are linked).

Not all traits are inherited in the simple dominance pattern that Mendel observed in his experiments. Some are inherited according to *codominance*. This is most obvious when each of two alleles possible for a gene contributes to a new, heterozygous phenotype that shows a pattern or combination of the two originals. This can be observed in petunias when a homozygous red-flowered plant reproduces with a homozygous white-flowered plant to produce offspring that all possess red-and-white patterned flowers.

Yet other traits are inherited according to *incomplete dominance*. In this case, each of the two alleles in a heterozygote again produce a new third phenotype. This offspring, however, appears to be a blended outcome of the other two alleles. Other flowers, like the snapdragon, demonstrate this inheritance pattern when a red-flowered parent and a white-flowered parent produce all pink-flowered offspring.

Exercise 6.2

If you can successfully work through the following questions, you should feel confident about being able to move on to the next section.

1. If an organism has one dominant allele and one recessive allele for a particular gene, that individual's genotype is best described as _____.
   a. heterozygous
   b. homozygous
   c. recessive
   d. dominant

2. Which of the following statements regarding genetic inheritance is true?
   a. One allele is always completely dominant over another.
   b. Only two phenotypes are possible for any trait.
   c. An organism will show either the dominant or the recessive phenotype for a given gene.
   d. A dominant allele will always mask the expression of a recessive allele when both are present.

3. Use the theme of the process and application of science to describe the concept of genetics as perceived by Mendel.

   _____

   _____

   _____

7

DNA and the Genetic Code

☑ DNA is the molecule that stores the genetic code for all forms of life.

☑ It is composed of a polymer of nucleotides; different sequences of nucleotides arranged into genes of varying lengths allow for the production of unique proteins and, thus, diverse genetic traits.

☑ Structurally, DNA is made up of two strands of complementary nucleotides. Recall that a nucleotide consists of a pentose sugar, a phosphate group, and a nitrogenous base.

☑ The strands within a DNA molecule run in opposite (antiparallel) directions. The two strands are held together by hydrogen bonds.

The Structure of DNA

Each strand within a DNA molecule is composed of a chain of covalently bonded nucleotides. The bond between nucleotides, called a *phosphodiester bond*, forms between the sugar of one nucleotide and the phosphate group of the other. This establishes a sugar-phosphate backbone that characterizes both DNA and RNA.

The sugar present within DNA is always a type called *deoxyribose*, and the four possible nitrogenous bases are *adenine* (A), *guanine* (G), *cytosine* (C), and *thymine* (T). Reading down the length of one of the strands allows the scientist to read the sequence of genes that the molecule of DNA holds by reading the specific sequence of nucleotides (really, of nitrogenous bases).

The two strands are held together through interactions that are strong enough to protect the genetic code and be less subject to mutation (change)

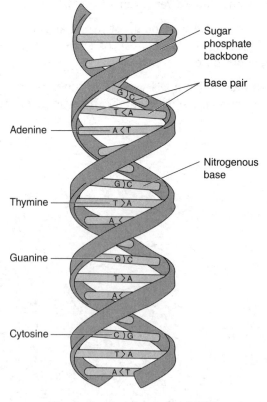

The molecular structure of DNA

but that are also weak enough to be pulled apart temporarily so the code can be read and copied. Hydrogen bonds are perfect for this purpose, so DNA has evolved such that one strand must be *complementary* to the other. This means the series of nitrogenous bases that run along one strand of DNA can form hydrogen bonds to the nitrogenous bases along the other strand. This can only be accomplished if there is some sort of coordination between the bases, which is what happens between complements. Within DNA, the A nucleotide is complementary to T and the C nucleotide is complementary to G.

The Central Dogma of Molecular Biology

Long after the fundamental principles of genetics were explained by Gregor Mendel, the molecular basis for heredity was finally understood. The molecular flow that explains the mechanisms for copying genetic information and the means by which traits are expressed from genes is called the central dogma of molecular biology.

Replication: DNA → DNA

Replication is the process by which the DNA strands are enzymatically separated by the breaking of the hydrogen bonds between complementary base pairs. Once the strands are effectively single, other enzymes work to read the sequence of nucleotides present in each strand and use that sequence as a template to construct a new, complementary strand. Free nucleotides are positioned across from their counterparts in the original DNA molecule, hydrogen bonds form, and the replication of the remaining sequence continues. When the entire initial strand is copied, replication is complete.

Transcription: DNA → RNA

Transcription is an essential component of the central dogma in that it allows a portion of the genetic code—carefully stored within the stable and predictable structure of the DNA molecule—to be transferred into a temporary RNA copy. The transcription process involves the DNA strand being separated temporarily so a complementary RNA transcript of a DNA gene can be made.

Translation: RNA → Protein

The RNA transcript produced during transcription is then used as the basis for building a protein during *translation*. The RNA transcript is read by a ribosome like a recipe: the transcript instructs the ribosome as to the correct sequence of amino acids to assemble into the desired protein.

Proteins as the Basis of Genetic Traits

Recall that proteins either are characterized as structural (like muscle fibers or keratin in skin cells) or are enzymatic and considered functional. Amazingly, proteins are diverse enough in structure and function to provide the molecular basis for all our traits. So DNA is really just a massive set of stored instructions for producing all the necessary proteins that a given organism needs to survive and carry out its life processes.

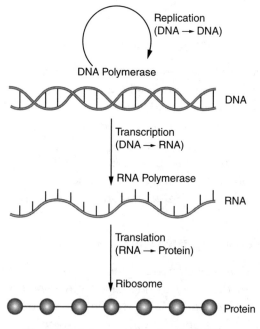

The central dogma of molecular biology

The central dogma can be summarized as the molecular flow of genetic information that is required for cells to grow and metabolize. Overall, the process can be represented by the following formula:

$$\circlearrowright DNA \rightarrow RNA \rightarrow protein$$

Here, the first arrow represents replication, the second represents transcription, and the third represents the process of translation. Collectively, the latter two arrows are terms for protein synthesis. Recall that the eukaryotic cell cycle included these metabolic processes during the G_1 and G_2 phases, which involve significant levels of protein synthesis as the cell metabolizes and grows. The S phase focuses on DNA replication after receiving a signal that the cell will initially divide into two.

Exercise 7.1

If you can successfully work through the following questions, you should feel confident about being able to move on to the next section.

1. Which of the following processes is responsible for building a copy of the genetic code before the cell divides?
   a. Transcription
   b. Protein synthesis
   c. Translation
   d. Replication

2. Which of the following statements regarding the central dogma of molecular biology is true?
   a. DNA is involved in transcription.
   b. DNA is involved in translation.
   c. RNA is involved in replication.
   d. All of the above are true.

3. Use the theme of continuity and variation to describe the concept of the genetic flow of information represented by the central dogma.

   _____

   _____

   _____

DNA Replication

Recall that DNA replication is the process by which a DNA molecule is copied. It occurs during the S phase of the cell cycle and indicates that the cell has received some feedback to encourage eventual cell division. The process of replication is carefully coordinated such that the copy of DNA possesses exactly the same sequence of nucleotides as the original template strand. If a mistake does occur, a mutation results.

Replication Step 1: Helicase Action

When in true double-stranded form, DNA is structured to protect the genetic code with the sequence of bases along either strand forming hydrogen bonds to the complementary sequence of bases along the other strand. To expose the code for copying, the two strands of DNA must be separated and each used to make a new, complementary copy. How are the two strands pulled apart? An enzyme, of course!

Helicase is that enzyme. It effectively spins the DNA molecule so fast that the hydrogen bonds break and the molecule unzips along that stretch. In a eukaryotic cell, multiple helicase enzymes work on an individual DNA molecule simultaneously, creating many replication forks to speed up the process. In a prokaryote, only a single, circular chromosome is present, so two replication forks start from a single origin of replication.

Replication Step 2: DNA Polymerase Action

Now that the DNA molecule has been opened at a replication fork, the next set of enzymes gets to work. These *DNA polymerases* bind to each exposed single strand of nucleotides, detect the type of base present at that location, and then use that information to position the complementary type of free DNA nucleotide directly across from the one that was read. A new set of hydrogen bonds then forms between the nucleotides.

Replication Step 3: Extension of New Strands

This process continues until (1) helicase has effectively opened the entire length of the DNA molecule being copied, and (2) DNA polymerase has effectively extended each exposed single strand to create the complementary copies. The major resource the cell needs to complete this process is a steady supply of free DNA nucleotides, which are available in the nucleus from the digestion of consumed and old, recycled nucleic acids.

Product: Two Identical DNA Strands

The two DNA strands that result from replication are identical, because each original strand was read and used as a recipe for making a new complementary strand. Because the mechanism of replication results in two DNA molecules that are each "half-old, half-new," it is called *semiconservative replication*.

During the eukaryotic cell cycle, recall that the S phase is the one during which DNA replication occurs in preparation for eventual mitosis.

A main goal of replication is continuity of the genetic code. This preserves instructions within all the cells of a multicellular organism or provides a means of quickly copying the parental genome before an asexual organism reproduces.

When there is a change in the genetic code, a mutation has occurred. After DNA replication, special proofreading enzymes scan the resultant copies to ensure that no mutations occurred during the process; any mis-

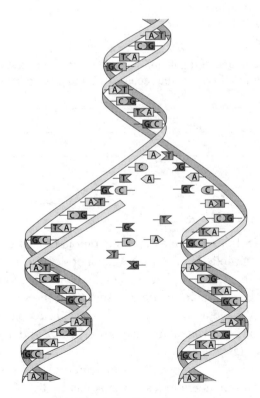

DNA replication overview

takes are edited accordingly. This process finds and corrects most errors, but other mutations can be introduced due to exposure to damaging chemicals or sometimes at random.

Exercise 7.2

If you can successfully work through the following questions, you should feel confident about being able to move on to the next section.

1. Which of the following molecules is responsible for providing the energy needed to expose the genetic code during replication?
   a. DNA polymerase
   b. Ligase
   c. RNA polymerase
   d. Helicase

2. Which of the following is (are) necessary for replication to occur?
   a. Free RNA nucleotides
   b. RNA polymerase
   c. Free DNA nucleotides
   d. Heat

3. Use the theme of regulation: feedback and response to describe the concept of DNA proofreading and repair that takes place following replication.

   _____

   _____

   _____

RNA Nucleotides

RNA nucleotides are an essential component to protein synthesis that is not used during replication. An RNA nucleotide, like a DNA one, is composed of a pentose sugar, a phosphate group, and a nitrogenous base. The differences between the two lie in the type of sugar and the selection of bases. The sugar in an RNA nucleotide is always *ribose* (a close cousin of the deoxyribose in DNA), and the nitrogenous bases are A, G, C, and *uracil* (U), which replaces thymine in DNA.

From a larger perspective, a molecule of RNA is always composed of a single strand of nucleotides, so it is easily distinguishable from the DNA double helix. That single strand is sometimes arranged in a simple, straight-

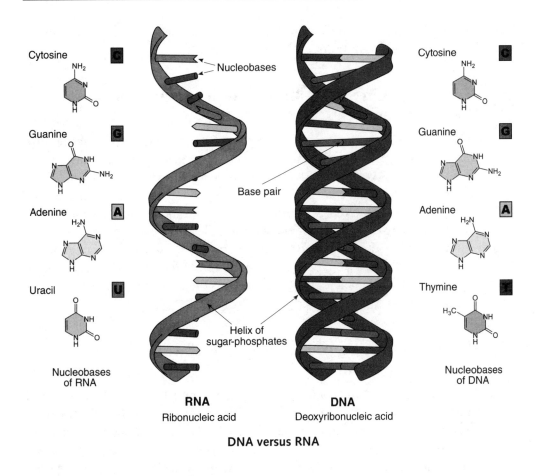

Cytosine

Guanine

Adenine

Uracil

Nucleobases
of RNA

Cytosine

Guanine

Adenine

Thymine

Nucleobases
of DNA

Nucleobases

Base pair

Helix of
sugar-phosphates

RNA
Ribonucleic acid

DNA
Deoxyribonucleic acid

DNA versus RNA

chain formation, as is present in *messenger RNA* (mRNA). In other cases, the single strand loops over on itself three times, using hydrogen bonds between complementary nucleotides·along the loop to hold it in place. This is characteristic of *transfer RNA* (tRNA). Finally, the single strand of RNA nucleotides can fold over on itself in many places, as is present in globular *ribosomal RNA* (rRNA). When rRNA associates with certain proteins that provide additional structure, the compound unit is called a ribosome, the organelle of protein synthesis.

Protein Synthesis

While DNA is responsible for the storage of genes, RNA is the molecule responsible for the actual expression of traits by the cell of an organism. The sequential processes of transcription and translation explain the mechanism for this expression of traits.

Transcription Step 1: Binding of RNA Polymerase to Promoter

Although transcription is similar to replication in that both processes require DNA in a single-stranded form to successfully read and copy it, transcription only requires a very short stretch of DNA to be exposed at a time. This means that no molecule such as helicase is required for transcription. Instead, the enzyme that will extend the new strand also breaks the hydrogen bonds between complementary nucleotides.

In this case, the enzyme RNA polymerase binds to a discrete region just before the start of a gene called the *promoter*. Only when RNA polymerase is bound at the promoter, along with a series of other molecules called *transcription factors*, does transcription fully begin.

Transcription Step 2: Assembly of the mRNA Transcript

RNA polymerase forces apart the hydrogen bond between the first set of nucleotides encountered in the gene following the promoter. The enzyme recognizes the particular nitrogenous base and then positions a complementary free RNA nucleotide across from it. Before any hydrogen bond can form, the enzyme moves down to the next set of nucleotides, again separating the double-stranded DNA, reading the nucleotide along one side, and then calling in a complementary nucleotide. The resultant molecule produced is called the *mRNA transcript*.

Transcription Step 3: Release at the Termination Signal

After the sequence of the entire gene is copied, the RNA polymerase enzyme reaches a point on the DNA molecule known as the *termination signal*. The enzyme has a very low affinity for DNA at this point, so it falls off the molecule. The mRNA transcript is complete, and it too separates from the DNA strand, leaving the last single-stranded stretches of exposed DNA to quickly re-form hydrogen bonds between complements.

Translation Step 1: Initiation of Translation

In eukaryotes, the mRNA transcript containing a complementary copy of a DNA gene now snakes out of the nucleus through a nuclear pore. It finds a ribosome in the cytoplasm and initiates translation, the production of the desired protein.

The ribosome pulls the mRNA transcript through its structure, reading a series of three nucleotides at a time. Each set of three bases, called a *codon*, codes for one of the 20 different amino acids that act as the building blocks of protein. The first codon found in any functional mRNA transcript is called a *start codon*, meaning that it is necessary to initiate the process.

When a ribosome reads a codon, it positions a complementary tRNA molecule in such a way as to encourage hydrogen bonding. It holds this tRNA in place until it can read the next codon and position the subsequent tRNA molecule appropriately.

Translation Step 2: Elongation of the Polypeptide

Why are tRNA molecules getting involved? Because of their three-looped structure, they are able to position a sequence of three nucleotides along one loop called the *anticodon*. Each anticodon is complementary to a specific codon possible in a mRNA transcript. On the opposite end of the tRNA molecule from the anticodon is the amino acid attachment site. Only the amino acid called for by the specific mRNA codon will attach to a tRNA molecule with the complementary anticodon.

During this elongation step, successive codons in the mRNA transcript are read, the appropriate tRNA molecules deliver the desired amino acids, and amino acids are added to the growing polypeptide.

Translation Step 3: Termination of Translation

The translation process proceeds until the ribosome reads the final codon in the transcript, called the *stop codon*. At this point, the ribosome loses its affinity for the mRNA molecule, all of the assembled components separate, and translation is complete.

Although the desired polypeptide has been synthesized, additional modification of the protein is often required before the cell can actually use it.

Protein synthesis is a complex process in eukaryotes, as it requires coordination of the transcription process in the nucleus and the translation pro-

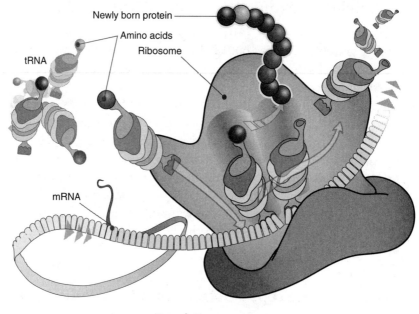

Newly born protein
Amino acids
Ribosome
tRNA
mRNA

Translation summary

cess in the cytoplasm. This physical separation of the two parts of protein synthesis also allows for built-in checkpoints to regulate the process. So if transcription occurs but the cell then receives feedback that it no longer needs the respective protein, then the transcript can be degraded before translation occurs. The cell thus saves precious time, energy, and other resources.

In prokaryotes, the lack of a nucleus means the transcription and translation processes are not physically separate, and both take place in the cytoplasm. Once the mRNA transcript produced from transcription begins to grow long and extend away from the DNA molecule, a ribosome immediately assembles around it and begins translation. This occurs while the transcript is still being built, so there is no opportunity for regulation at this point.

 It is easy to confuse enzymes during these processes. Remember that DNA polymerase is only required when a new molecule of DNA is being built—and then only during replication. RNA polymerase is similarly required to build a new polymer of RNA and is used only during transcription.

It is important to remember that transcription and translation must occur in sequence for protein synthesis to take place. It is helpful to understand that the word *transcription* comes from roots meaning "to write across." This suggests the process of reading a code in one language (initially a sequence of DNA nucleotides) and transferring it to a different, complementary code in another language (a sequence of RNA nucleotides).

Translation takes the RNA code and translates it into yet another code, this time a sequence of amino acids in a polypeptide.

Exercise 7.3

If you can successfully work through the following questions, you should feel confident about being able to move on to the next section.

1. Which of the following is (are) *not* required for protein synthesis to occur?
   a. RNA polymerase
   b. Free RNA nucleotides
   c. Helicase
   d. Both a and b

2. To which aspect of cellular metabolism is the process of translation most closely linked?
   a. Anabolism
   b. Dehydration synthesis
   c. Catabolism
   d. Both a and b

3. Use the theme of the relationship between structure and function to describe the significance of the three different forms of RNA involved in protein synthesis.

   _____

   _____

   _____

8

Human Genetics and Biotechnologies

☑ As humans, we inherit many of our genetic traits in much the same way that other organisms do. There are also some complex forms of inheritance that make heredity of certain genes more difficult to predict.

☑ Biological sex is determined genetically by the presence or absence of the Y chromosome; females lack a Y chromosome and are characterized as XX; males possess a Y chromosome and are characterized as XY.

☑ A good number of our traits are *monogenic*, meaning that their expression is affected by the presence of a single gene. Monogenic traits can be characterized as autosomal dominant or recessive, or instead as X-linked autosomal or recessive.

☑ Others genes are described as *polygenic*, meaning that the final phenotype is influenced by the interaction of three or more different genes. Polygenic genes are much more challenging to predict from generation to generation because of the many possible combinations of alleles and genotypes.

The Human Genome

The human genome consists of 23 paired chromosomes, one chromosome in each pair inherited from each parent. Many genes are held on each individual chromosome, and the final pair of chromosomes is even responsible for determining the human's biological sex.

Karyotypes

A *karyotype* is a laboratory procedure that produces an assemblage of diploid, eukaryotic chromosomes from a cell that is arrested during metaphase in mitosis. An image of these chromosomes is taken to preserve the data for future analysis. The chromosomes are then spliced out individually and paired up with their homologue. For practical purposes, they are arranged from largest chromosome (pair #1) to smallest chromosome (varies depending on the species; in humans, it is pair #22), with the distinct sex chromosomes positioned at the very end (regardless of size).

Autosomes

An autosome is any chromosome within a karyotype (or the natural genome of a cell) that does not have any involvement in the determination of biological sex. Most chromosomes within a typical karyotype are autosomes, carrying the vast majority of genes necessary for the organism's survival. Humans have 22 pairs of autosomes in their genome; some species have a few pairs (for example, fruit flies have 4), while others have many (for instance, some ferns possess more than 1,000).

Sex Chromosomes

Sex chromosomes are the pair of chromosomes that determine biological sex, and they often come in two forms called X and Y. The pattern that specifically determines male or female, however, differs drastically depending on the species. As humans, we might be accustomed to the idea that males have a Y chromosome (the XY pattern), while females do not (the XX pattern). While this is true for mammals, fruit flies, and some other species, it is not so for all of them. In birds, males have two homologous sex chromosomes and females do not. In many insects, there is no Y chromosome at all; an unfertilized X egg produces a male offspring, while only a fertilized XX egg produces a female.

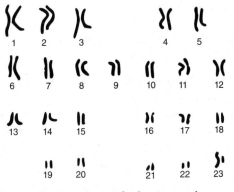

The karyotype of a human male

Sex-Linked Traits

Given that the composition of sex chromosomes distinguishes male from female in many species and that these chromosomes also hold genes for other traits that seemingly have nothing to do with biological sex, some interesting outcomes result. For example, a human male has only one X chromosome and therefore only one copy of a certain gene required for proper blood clotting. A relatively common mutation exists at that location, and the result of possessing only the mutated, recessive alleles is the X-linked condition of *hemophilia*. (It is commonly referred to as the bleeder's disease due to the lack of clotting proteins.)

Because males have only one copy of this gene, a male who possesses a single hemophilia allele will inherit hemophilia (even though the allele is recessive). A female has the protection of two X chromosomes, so if she possesses only one mutated hemophilia allele, it follows that the corresponding allele on the other X-chromosome is normal and dominant. She therefore does not express the hemophilia condition but is a genetic carrier and can pass it on to her offspring if they inherit that particular X chromosome.

The concepts of sex chromosomes and sex cells (or gametes) are often and easily confused. It is important to remember that eukaryotic organisms are expected to have a full set of chromosomes—autosomes and sex chromosomes alike—in the nucleus of every cell. In a diploid cell, the nucleus possesses two distinct sets of the genome, one inherited from each parent. In a gamete, the cell is expected to be haploid and possess only one set of chromosomes—one of each type but usually a mix from both parents due to genetic recombination.

Exercise 8.1

If you can successfully work through the following questions, you should feel confident about being able to move on to the next section.

1. Which of the following is true of autosomes but not of sex chromosomes?
    a. They determine biological sex.
    b. They contain several genes known to cause X-linked disorders like hemophilia.
    c. They house the majority of genes found in an organism's genome.
    d. They are not truly homologous in human males.

2. Which of the following should be true for a human karyotype from a somatic cell?
    a. There are 44 pairs of autosomes.
    b. There are two pairs of sex chromosomes.
    c. There is one pair of sex chromosomes at position #22.
    d. There are 44 total autosomes and two sex chromosomes.

3. Use the theme of the process and application of science to describe the concept of a karyotype.

    _____

    _____

    _____

Other Complex Patterns of Human Inheritance

Not all genetic traits found in humans can be explained using basic Mendelian principles. Often, more complex patterns of heredity and interactions between genes are responsible for creating novel genetic traits.

Polygenic Traits

Many traits familiar to humans are inherited in a polygenic fashion; that is, the final phenotype results from the interaction of many individual genes. These include traits like height, skin color, eye color, and others in which the phenotypic range does not comprise discrete characteristics but instead lies on a spectrum.

Multiple-Allele Traits

Other traits result from the transcription and translation of a single gene, but the gene actually contains more than the expected two alleles. Traits that result from such an inheritance pattern are therefore called *multiple-allele traits*. The most common such trait in humans is the one that produces different blood types (like A, B, AB, and O). In fact, the reason so many phenotypes are possible is that the gene has multiple alleles from which to work.

Incomplete Dominance

Incomplete dominance in humans is similar to the incomplete dominance discussed earlier in Chapter 6. Recall that incomplete dominance describes the inheritance observed when neither allele possible for the gene is dominant over the other, and a new phenotype is produced that is intermediate between the two phenotypes that result from the homozygous condition. In humans, this is observed in the familiar condition of elevated blood cholesterol levels. While this condition can result from a diet that is high in cholesterol, it often results from a defective gene. When the homozygous condition is present, a more severe form of high cholesterol is also present. When the heterozygous genotype is found, an individual has an intermediate form of elevated cholesterol.

Sex-Influenced Traits

While a sex-influenced trait indirectly involves the sex chromosomes, the genes that code for the condition are not housed on the sex chromosomes as those for sex-linked traits are. Instead, the genes for these traits are *autosomal* (found on autosomes), but their phenotypic expression is strongly influenced by the presence or absence of certain sex hormones.

In humans, the condition known as *pattern baldness* results from sex-influenced inheritance. The baldness allele is expressed as dominant in males due to the interaction with the male sex hormone testosterone. Females lack particularly high levels of testosterone, and the expression of the baldness gene is recessive. This means it is easier for males to show a sex-influenced condition like pattern baldness just as it is easier for them to express an X-linked trait like hemophilia, but the reasons for each are quite different.

Single-Allele Traits

Because most human genetic disorders are the result of a mutation that prevents a usual protein from functioning normally, they are predominantly recessive in nature. Every so often a mutation creates a new protein where none was produced before (or an excess of a typical protein), and a dominant disorder is then produced.

One of the most feared single-allele disorders in humans is *Huntington's disease*. This is due not only to its allele being dominant but also to the late onset of the disease (symptoms don't appear until age 50 or later); it is fatal.

Examples of monogenic traits include benign ones, such as the presence or absence of a widow's peak along the hairline of the scalp or earlobes that are free-hanging or attached directly to the head. Such traits also include some serious and potentially life-threatening genetic disorders and diseases like cystic fibrosis and dwarfism.

Sex-linked traits and sex-influenced traits are easily confused. It is important to remember that genes for sex-linked traits are physically linked to the sex chromosomes, so the genes are actually found on either the X or the Y chromosome. Sex-influenced traits are autosomal and thus not sex-linked, but their expression is influenced by the presence or absence of sex hormones. However, the genes for the hormones themselves are found on the sex chromosomes, so the link to sex-influenced traits is indirect.

Genetic Mutations

As previously discussed, many human genetic disorders are linked to spontaneously created or inherited mutations. Some are restricted to the confines of a single gene and are thus called *gene mutations* (or *point mutations*). If an original nucleotide is replaced by another, then a *substitution mutation* has occurred. If instead an extra nucleotide is squeezed into the original sequence or a nucleotide is lost, then either an insertion or a deletion has taken place. Recall that mRNA transcripts—and thus the DNA from which they were copied—are read in a series of three nucleotides at a time. This means that when an insertion or a deletion form of gene mutation occurs, the entire reading frame is changed from that point on. These are thus called *frameshift mutations*; while any can be harmful, those that occur near the beginning of a gene affect more amino acids in the resultant protein.

Other mutations are much bigger in scope than gene mutations and involve large pieces of a chromosome or the entire chromosome itself. These mutations also occur in some typical forms. *Deletion mutations* result when

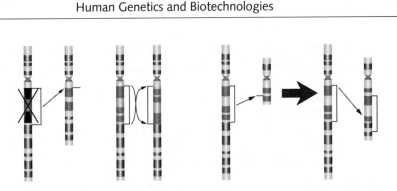

Three chromosome mutations: deletion, inversion, and translocation

a piece of a chromosome breaks off and is lost, whereas *inversion mutations* take place when a fragment reinserts itself at the breakage point but in the opposite direction. On other occasions, a chromosome fragment breaks off and inserts itself into another, nonhomologous chromosome; this is called *translocation*.

The nondisjunction type of chromosome mutation is the largest in scope and the one that is responsible for Down syndrome. A *nondisjunction mutation* occurs during meiosis when homologous chromosomes fail to separate appropriately. The result is a gamete with either too many or too few chromosomes. *Down syndrome*, the most common nondisjunction disorder, is also known as trisomy 21 due to the three chromosomes that are observed in a karyotype of an affected individual. Other disorders are monosomal, because they result from the inheritance of a gamete that was short a chromosome.

Exercise 8.2

If you can successfully work through the following questions, you should feel confident about being able to move on to the next section.

1. Which of the following human genetic disorders is *incorrectly* matched with its form of inheritance?
   a. Incomplete dominance—high cholesterol
   b. Polygenic inheritance—skin color
   c. Single-allele trait—ABO blood type
   d. Sex-influenced condition—pattern baldness

2. Which of the following statements regarding human genetic expression is *false*?
   a. A nondisjunction mutation produces gametes with an incorrect number of chromosomes.
   b. A condition that results from incomplete dominance is one that involves a hetcrozygous gcnotypc.
   c. All human genetic disorders are recessive in nature.
   d. Gene mutations only affect one gene at a time, while chromosomal mutations can affect multiple genes simultaneously.

3. Use the theme of the relationship between structure and function to describe the concept of chromosomal mutations.

   _____

   _____

   _____

DNA Fingerprinting

Recently, technology and research in modern molecular genetics have allowed for DNA to be utilized for many practical purposes. Now familiar to the layperson through the explosion of forensics and crime-solving references in popular culture, DNA fingerprinting represents one such application.

Restriction Enzymes

Restriction enzymes occur naturally in bacterial cells as a protective mechanism against viral infection; they function by reading DNA molecules and making predictable cuts along specific sequences. Researchers use restriction enzymes in the first step of DNA fingerprinting to cut different DNA samples that are under comparison. If this is being done as part of a forensic analysis and crime scene DNA is being compared with that of three potential suspects, only the actual perpetrator will have DNA fragments that are identical to those produced from the crime scene sample. This is because restriction enzymes make cuts only along very specific sequences, and overall DNA sequences within a genome are identical only in the same individual or among identical twins.

Polymerase Chain Reaction

Polymerase chain reaction (PCR) should be thought of as a DNA photocopy machine. It simulates many of the steps of DNA replication (discussed in detail in Chapter 7), but it does so in such a sophisticated, precise way in the controlled setting of a research lab that billions of copies of DNA can be synthesized in a matter of hours. In a natural setting, replication occurs only when future cell division is anticipated as a controlled part of the cell cycle. A key way that PCR is distinct from replication is that PCR uses heat to separate hydrogen bonds and does not employ an enzyme like helicase. Once the DNA has been separated into two strands, primers, DNA polymerase enzymes, and free DNA nucleotides are used to build two new complementary strands, one from each template. This cycle is repeated as many times as necessary to produce all of the copies needed for thorough analysis by all interested parties and for macroscopic visualization of each fragment in the eventual fingerprint.

Gel Electrophoresis

Each sample of copied DNA fragments is carefully pipetted into a tiny well carved out of a gelatinous substance called an *agarose gel*. This situates the DNA fragments nicely within the thickness of the gel. Once all of the samples are loaded, charges are applied to the apparatus using electrodes. A negative electrode is placed near the well where the DNA samples were loaded, while a positive electrode is situated at the opposite end of the gel.

Recall that DNA is negatively charged due to the presence of many phosphate groups. The negative electrode pushes on the DNA fragments, while the positive charge pulls them. Ultimately, the DNA fragments are separated out by size, with the smallest moving fastest and migrating closest to the end of the gel. A specific pattern results that should be unique to a given individual if enough specific sites along the genome are used in the analysis.

DNA Fingerprint Analysis

The results of the gel electrophoresis are generally stained and/or preserved via an imaging technique for ongoing analysis. This creates the DNA fingerprint, the final pattern of bands that results for each original sample of DNA under comparison. Remember that only identical twins will have the same DNA fingerprint, so if two of the samples produce identical banding patterns in the DNA fingerprint, that constitutes an identity match!

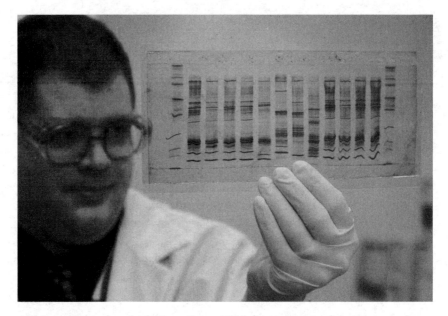

A U.S. customs chemist interpreting a DNA fingerprint to determine an item's origin

Cloning

Many other applications of DNA technologies are widely used and being researched. Cloning is once such application, and it comes in two forms. *Reproductive cloning* involves producing a new organism from the genome of an organism that has already existed. This is accomplished by removing an egg cell from one organism, replacing its nucleus with that of a diploid somatic cell from the organism you want to clone, and then stimulating the "fertilized" egg to divide and develop into an embryo. When the embryo achieves a certain stable state called the *morula*, it is transferred into the uterus of a surrogate mother who nourishes and protects the embryo during gestation and through birth.

Another kind of cloning that is far less controversial is *gene cloning*. This process uses a bacterial colony as a manufacturing plant for a desired human protein. It is accomplished by removing a healthy donor gene from human DNA (using restriction enzymes to cleave out the gene) and inserting the donor gene into a notch created in a bacterial *plasmid*, a small, circular piece of DNA possessed by bacteria in addition to their main chromosome. Because the notch in the plasmid is created using the same restriction enzyme that spliced out the human gene, the ends fit together perfectly via hydrogen bonding. Next, the bacterium is encouraged to reproduce using binary fission and thus create a population of cloned bacteria that all contain the instructions for the human protein. Then the bacteria are encouraged to undergo transcription and translation of the human gene and create protein. This process is widely used to manufacture insulin, the protein-based hormone that is deficient and/or dysfunctional in individuals with diabetes.

Exercise 8.3

If you can successfully work through the following questions, you should feel confident about being able to move on to the next section.

1. Which of the following biotechnologies is *incorrectly* matched with a use?
   a. Gene cloning—making a cloned mouse from an existing mouse
   b. Gel electrophoresis—separating DNA fragments by size
   c. DNA fingerprinting—establishing genetic identity for forensics
   d. PCR—copying DNA in mass quantities

2. Restriction enzymes are useful in _____.
   a. gene cloning
   b. DNA fingerprinting
   c. joining together different pieces of DNA
   d. all of the above

3. Use the theme of continuity and variation to describe the concept of reproductive cloning.

   _____

   _____

   _____

9

The Origins of Life on Earth

☑ Before the Scientific Revolution, people believed that much new life could be explained by *spontaneous generation*. Many also believed that a "vital force" was carried in the air that could spark life from nonlife.

☑ During the Scientific Revolution, several scientists offered up new hypotheses regarding the creation of new life. Based on 200 years of evidence, a consensus was finally reached: spontaneous generation was refuted, and a new concept called *biogenesis* was supported.

☑ Ongoing research and the discovery of fossils helped scientists to better understand the origins of the Earth and the life it has supported and will continue to sustain over time.

Early Experiments in the Origins of Life

While most people at the time believed in spontaneous generation and/or creationism, a few early scientists were hard at work trying to uncover the mechanisms responsible for the origins of life.

Francisco Redi

In the mid-1600s, the Italian scientist Francisco Redi observed fly development and began to question the commonly held assumption that maggots generated spontaneously from dead meat. He conducted an experiment to test his hypothesis that flies laid eggs on the meat and the eggs developed into maggots, a form of fly larva. While the experimental group, one that contained meat in a jar covered with netting, did not produce any maggots, the uncovered meat in the control group did. Redi's work definitively disproved the previously held belief in the spontaneous generation of maggots, but it did not disprove the general notion entirely. That would take 200 years!

Lazzaro Spallanzani

Nearly a century later, another Italian scientist, Lazzaro Spallanzani, attempted another experiment to disprove spontaneous generation, this time involving microbes. He knew that microorganisms could easily contaminate food, so he chose to experiment on an organic broth. Spallanzani first boiled the broth in two glass flasks to kill off any existing microbes and then sealed one to make the experimental group. The other flask was left open and exposed to microbes in the air; this was the control group. Not surprisingly, no growth was observed in the broth in the sealed experimental group, but the broth in the control group teemed with microbial life.

While Spallanzani and many others in the scientific community were satisfied that the results further disproved spontaneous generation, critics said the sealed flask kept out the vital force in the air that was required to spontaneously generate life from nonlife. Another century would pass until the concept was tested one final time.

Louis Pasteur

Well-known French scientist Louis Pasteur attempted to disprove spontaneous generation once and for all in the mid-1800s. He devised an ingenious flask that had a long, thin, curved neck. This neck would allow the air (and

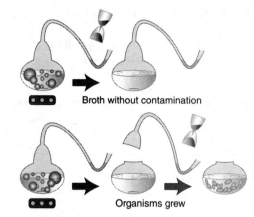

Pasteur's experiment and the support of biogenesis

thus any vital force) to mix with the contents of the flask but would not allow any microorganisms from the air to enter, due to its long, convoluted pathway. Pasteur boiled organic broth in the flask to decontaminate it and then let it sit for one year. No growth occurred in the broth during that entire time. Pasteur then broke the curved part of the neck off the flask to create a standard straight neck through which microorganisms could fall from the air into the broth. The broth became cloudy with microbial life within a day.

Finally, the concept of spontaneous generation was disproved and a new principle, biogenesis, was embraced. *Biogenesis* simply states that only life can generate life. But that raises the question: how did the first life on Earth begin?

Exercise 9.1

If you can successfully work through the following questions, you should feel confident about being able to move on to the next section.

1. Which of the following is true regarding spontaneous generation?
    a. It is responsible for the origin of some simple life.
    b. It works only when a vital force is present.
    c. It was finally disproved by Spallanzani.
    d. It was eventually replaced by the concept of biogenesis.

2. Which of the following best describes Pasteur's experimental group?
   a. He didn't have one.
   b. It was the flask with the original curved neck.
   c. It was the flask with the broken straight neck.
   d. It was not adequately described.

3. Use the theme of continuity and variation to describe the concept of biogenesis.

   _____

   _____

   _____

From an Inorganic World to Cellular Life

Early Earth was tumultuous and inhospitable to life, characterized by high temperatures, ongoing volcanic activity, and bombardment by meteorites and other space debris. Eventual and significant cooling over time made the Earth habitable with the condensation of bodies of liquid water from the atmosphere.

Primordial Soup

After cooling, Earth consisted of large bodies of liquid water covering much of the terrestrial (rock) formations. The water would have acted as a solvent for many of the simple, inorganic molecules present at the time. Others would have been found in the mixture of gases that remained in the atmosphere.

Organic Molecules

If conditions are optimal, organic molecules can form from inorganic ones—such as hydrogen gas (H_2), methane (CH_4), ammonia (NH_3), and water (H_2O)—without the coordinated metabolic activity of organisms. The synthesis of these more complex molecules required a large input of energy, as in lightning or ultraviolet bombardment. In 1924, biochemist Alexander Oparin proposed this idea but never determined how to test it scientifically. Finally, in 1953, graduate student Stanley Miller, working under the supervision of chemistry professor Harold Urey, devised the Miller-Urey apparatus and experiment. The two demonstrated that simple organic molecules like amino acids, ATP, and nucleotides can be produced from a gaseous mixture

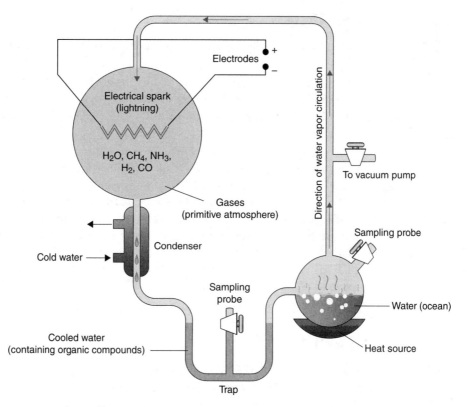

The Miller-Urey experiment and the origin of organic molecules

of inorganic molecules that are heated, sparked with energy via electrodes, and then condensed.

Protocells

When a variety of simple organic molecules interact in a solution, certain assemblages commonly form (much like soap bubbles). Small, spherical structures with a protein membrane that spontaneously form under certain conditions are called *microspheres*. Additional spherical structures, called *coacervates*, are composed of other substances, including lipids and sugars.

When some of these spontaneous structures actually formed early in the Earth's history, they trapped RNA molecules, ATP, and other helpful substances. These likely gave rise to *protocells*, the multiple attempts at early cellular life that were largely unsuccessful.

First True Cells

Around 3.8 billion years ago (bya), true cells formed from protocells. These cells were likely *heterotrophic*, relying on a solution of organic molecules for energy. Due to the lack of oxygen gas at the time, these cells are also presumed to have been *anaerobic*. Organisms living today that most closely resemble these first life-forms belong to the Archaebacteria kingdom.

Don't worry if remembering the sequence of events in the origin of the first true cells seems overwhelming. It is important to remember that one level easily influences the next as the structure becomes closer and closer to one that can actually be classified as living. That is, inorganic molecules were needed to build organic ones, then larger, cell-like spontaneous assemblages formed due to the interactions of the organic molecules in a particular environment. Only protocell assemblages that contained some form of nucleic acid, ATP for energy, and so on were likely fit and potentially able to account for some of the first true cells.

Exercise 9.2

If you can successfully work through the following questions, you should feel confident about being able to move on to the next section.

1. Which of the following statements is *not* true of the Miller-Urey experiment?
   a. Some of the products were amino acids and nucleotides.
   b. Some of the reactants were water and oxygen.
   c. Electrodes were used to simulate a major energy event.
   d. The reactants simulated a primordial soup.

2. Which of the following structures are *not* spontaneously generating or significant in understanding the origins of the first life on Earth?
   a. Protocells
   b. Coacervates
   c. Microspheres
   d. Archaebacteria

3. Use the theme of continuity and variation to describe the concept of the origin of the first true cells.

   _____

   _____

   _____

Early Events in Evolutionary History

The period of time between the creation of the Earth more than 4.5 bya and the origin of the first prokaryotic cells more than 3.8 bya allowed for many spontaneous "attempts" at life before producing some that actually worked in a particular environment. Prokaryotes flourished on the Earth as the lone inhabitants for at least 2 billion years, along the way diversifying in structure and function as evolution occurred.

Anaerobic Prokaryotes

Anaerobic prokaryotes dominated the Earth and eventually would have consumed all of the organic molecules present if left to their own devices. Levels of carbon dioxide (CO_2) began to increase significantly as a by-product of anaerobic metabolism, and conditions became suitable for a new type of life to evolve.

Autotrophic Prokaryotes

Chemoautotrophic prokaryotes likely evolved first. These were able not only to tolerate but also to thrive in the conditions present on early Earth. Using CO_2 and producing simple organic molecules, these prokaryotes relied on energy from inorganic molecules in their surroundings.

Only after a significant period, approximately 3 bya, did *photoautotrophic prokaryotes* evolve that were capable of producing glucose by means of photosynthesis. These autotrophic prokaryotes, however, changed the fate of the world forever, because the by-product of their metabolic pathway was oxygen gas (O_2).

Aerobic Prokaryotes

Significant levels of O_2 began to accumulate over time. It was still toxic to many anaerobic organisms at the time, but some prokaryotes developed a way to use the molecule as energy. Aerobic respiration and the first aerobic prokaryotes then evolved some 2.5 bya.

Eukaryotic Cells

Many prokaryotes became more effective and efficient metabolically. This, in turn, supported the growth of more complex structures. First, eukaryotic

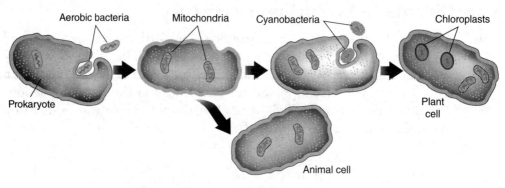

Endosymbiosis and the origin of eukaryotes

cells evolved approximately 1.5–2 bya through a process called *endosymbiosis*. Imagine that a large, heterotrophic bacterium engulfs a smaller prokaryote but cannot break it down. By chance, the ingested prokaryote is a specialist at cellular respiration and continues to function while inside the other cell. If, over time, the two cells become completely interdependent on each other, the eukaryotic cell evolves. The internal prokaryote would then become reclassified as a mitochondrion. This same process occurred to create the chloroplast, but the ingested prokaryote was a specialist at photosynthesis.

Multicellular Life

Eukaryotes were quite successful in their new, more complex form, but even more complex structures evolved over time. First, eukaryotic cells began living together in groups called *colonies*. In a colony, each individual organism contributes something to the whole but gains more in return. Some colonies became completely interdependent over time, and true multicellular life was formed for the first time approximately 1 mya (million years ago).

 Chemoautotrophs and photoautotrophs are easily mixed up. It is important to remember that the prefix in each word describes the energy source on which the autotroph relies to build its own food. Both autotrophs use CO_2 as the inorganic building block for their organic products, but the specific product generated depends on the type of metabolism employed.

Exercise 9.3

If you can successfully work through the following questions, you should feel confident about being able to move on to the next section.

1. Which of the following characteristics would *not* accurately describe the first cells?
   a. Prokaryotic
   b. Colonial
   c. Anaerobic
   d. Heterotrophic

2. Which of the following significant evolutionary events occurred after all the others?
   a. Endosymbiosis
   b. Aerobic prokaryotes
   c. Photosynthetic prokaryotes
   d. Multicellularity

3. Use the theme of interdependence in nature to describe the concept of endosymbiosis.

   _____

   _____

   _____

10

Evolution and Adaptation of Life

☑ Although Louis Pasteur's work disproved spontaneous generation and fully cemented the notion of biogenesis, most people still believed in the idea of the *immutability of species*. Largely due to the literal belief in religious doctrine, people assumed that each species was created in its current form and wasn't capable of changing into new species.

☑ All of that began to change when Charles Darwin shared his theory of evolution by means of natural selection with the world.

☑ For a long time, fossils and biogeography provided the basis for most evolutionary evidence. Eventually, other techniques, including those involving comparing molecules or embryonic development, supported earlier evidence.

☑ Currently, evolution is understood on a small scope by examining allele frequencies in a population and on a large scope by the creation of new species from ancestral ones.

Evolution by Means of Natural Selection

Charles Darwin's voyage as a naturalist on the HMS *Beagle* from 1831 to 1836 allowed him to circumnavigate the globe and encounter all sorts of organisms in myriad ecosystems. He collected vast quantities of plant and animal specimens, recorded countless observations in his log, and even sketched the first phylogenetic tree that showed relatedness and divergence of organisms. It took Darwin more than 20 years, however, to refine his ideas and eventually publish his best-known work, *On the Origin of Species by Means of Natural Selection*.

Genetic Variation

Evolution ultimately depends on the existence of genetic variation within a population. Recall that genetic variation is introduced during meiosis among sexually reproducing organisms through the processes of crossing over and independent assortment. Genetic variation is also increased by the random fertilization of zygotes and the spontaneous mutation events that can occur.

Overreproduction

The creation of more offspring than are capable of surviving to maturity in a given environment is typical in nature. Consider the runt of a litter that wouldn't survive without human intervention, the egg in a clutch that won't hatch (and might provide sustenance to its siblings), or the thousands of fertilized fish eggs that are consumed by other fish that are higher up in the food chain.

Struggle for Survival

Necessary resources are usually limited in a natural environment. These limiting factors—environmental factors that affect the survival and/or reproduction of a species—play an important role in evolution. They create an environment in which individual organisms compete with others for resources that affect their ability to survive and/or reproduce.

Differential Survival and Reproduction

Only organisms that possess a given collection of traits that makes them fit for survival in the environment will live. They will have the tendency to outcompete other organisms whose traits are less fit evolutionarily when resources are limited. Those organisms that have certain genetic traits that enable them to survive longer and more easily will also tend to produce more offspring. Thus, the next generation will disproportionately resemble those organisms with the fittest genetic traits. This is evolution in action.

Darwin's first phylogenetic tree sketched on the HMS *Beagle*

Although Darwin is credited with developing the theory of evolution by natural selection, it does not mean he was the first to conceptualize evolution or the only person to think of natural selection. His grandfather, Erasmus Darwin, believed in evolution, as did many others along the way, but Charles Darwin first developed the mechanism of natural selection to explain how evolution happens.

In fact, another biologist and naturalist, Alfred Wallace, came up with an idea similar to Darwin's before Darwin actually published. Many people don't realize it, but Darwin and Wallace are jointly credited with developing the theory of evolution by means of natural selection.

 # Exercise 10.1

If you can successfully work through the following questions, you should feel confident about being able to move on to the next section.

1. Which of the following is *not* an essential component of Darwin's theory of evolution by means of natural selection?
   a. Overpopulation
   b. Genetic variation
   c. Predation
   d. Differential reproduction

2. Which of the following could be a *biotic* (living) limiting factor in an aquatic ecosystem?
   a. Availability of sunlight
   b. Access to water
   c. Levels of parasitic fungus present
   d. Low temperature

3. Use the theme of continuity and variation to describe the concept of evolution in a population.

   _____

   _____

   _____

Evidence for Evolutionary Theory

If asked to provide evidence for evolution, one might provide the fossil record as a wonderful example, and for good reason. Most of us conjure up images of imposing dinosaur fossils at a natural history museum, but fossils can provide an account of past life for only a small sampling of Earth's biodiverse organisms.

The Fossil Record

The fossil record is extensive and provides a thorough foundation for understanding the evolutionary history of life on Earth. A record of the planet in geologic time is preserved to a certain extent in the various strata (or layers) present in sedimentary rock. The farther down you dig into the strata, the older the layers and any fossils within them typically are; this describes the *law of superposition*. The concept of determining a comparative fossil age based on superposition in strata is called *relative dating*.

Fossils and the rocks in which they are embedded can also be tested for the presence of certain radioactive isotopes (such as carbon 14). Because a radioactive isotope decays at a predictable rate over time, the proportion of it to the more common, stable form allows biologists to determine the absolute age of a specimen. This provides the fossil's actual age in years.

Comparative Anatomy (Structural Homologies)

Not all species throughout history have been able to fossilize. Some are not constructed of hard enough body parts to preserve well, and others die in a region that does not produce sedimentary rock. Still, biologists want to understand evolutionary relationships between currently living organisms, so *comparative anatomy* is useful for such cases and can support other types of evolutionary evidence.

Homologous structures are present in closely related species due to their shared genes that were inherited from a recent common ancestor. Such structures always share some underlying similarity, even when the current functions in related species are quite different. A classic example is the forelimb of mammals: a wing of a bat (used for flight), a flipper of a dolphin (used to swim), and a forearm of a chimpanzee (used to climb and manipulate tools). While these three structures seem strikingly different superficially, the underlying bone structure is actually quite similar.

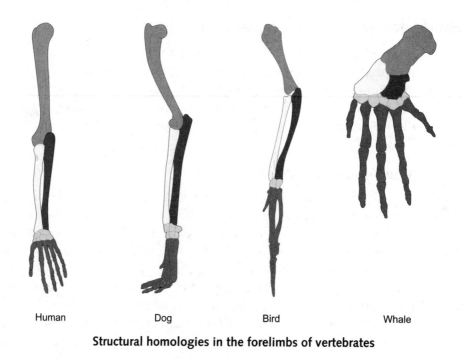

Human   Dog   Bird   Whale

Structural homologies in the forelimbs of vertebrates

Analogous structures are those that may look similar superficially, but the underlying structure is quite different and the genes responsible for the structure were not inherited from a recent common ancestor. Instead, the adaptations evolved independently in the two lineages, likely due to similar environmental pressures and limiting factors. Analogous structures do not provide evidence for relatedness.

Finally, *vestigial structures* are those that persist in a diminished, non-functional form in a modern species. Examples of vestigial structures include a human's wisdom teeth and a whale's pelvis. A close relative of a species with a vestigial structure may still retain a fully functional form of the structure. For example, a whale's pelvis was derived from the pelvis of a terrestrial piglike animal that inhabited the coasts. Modern pigs and other terrestrial mammals retain the functional pelvis necessary for coordinating movement of the lower limbs.

Comparative Embryology

With technological advances and the uncovering of the structure of DNA in the early 1950s, it was discovered that closely related species tend to share patterns of embryological development. This is due to a larger proportion of shared genes that are responsible for the early changes that take place in an embryo after fertilization to establish the body plan and develop the organ systems.

Comparative Biochemistry

Even more recently, over the past few decades, research on the structures and functions of nucleic acids and proteins has provided valuable insight into evolutionary history and relatedness among organisms. Given that DNA stores the genetic code in sequences of nucleotides called genes, and proteins result from decoding a gene into a sequence of amino acids within a polypeptide, comparing monomer sequences within either molecule can provide invaluable information regarding relatedness.

Traits that increase in frequency in a population because they present an increase in fitness are called *adaptations*. Just as a population (rather than an individual) evolves, so too does it adapt. An individual contributes to this process by acting as the unit by which a unique collection of genes is tested out in the environment, but evolution and adaptation are truly population-level characteristics.

When an individual organism changes in the short term as a response to a significant environmental shift, this is called *acclimatization*. It is observed when the human body makes more red blood cells in response to a decrease in oxygen levels, as when hiking at high altitudes.

Exercise 10.2

If you can successfully work through the following questions, you should feel confident about being able to move on to the next section.

1. Which of the following is most useful in estimating the age of a fossil in years?
   a. The law of superposition
   b. The fossil's relative age
   c. The presence of a radioactive isotope
   d. All of the above

2. Which of the following would *not* be useful in providing evidence for relatedness between species?
   a. Homologous structures
   b. Analogous structures
   c. Vestigial structures
   d. Comparative embryology

3. Use the theme of the relationship between structure and function to describe the concept of homologous structures.

   _____

   _____

   _____

From Microevolution to Macroevolution

Evolution as conjured up by Charles Darwin is in modern times best understood using one of two lenses. *Microevolution* explains evolution on the smallest scale, as a population of organisms passes on traits to subsequent generations. *Macroevolution* approaches evolution from the largest scope, focusing on the means by which new traits accumulate to create novel species and on understanding relatedness of all organisms over evolutionary time.

Gene Pool

It is important to realize that not all individuals in a population are capable of reproducing in their current state. If an organism cannot reproduce, then it cannot directly contribute to evolution in its population. Only individuals in a population that are capable of contributing their genes to the next generation are considered part of the *gene pool*.

Frequencies of alleles present for a given gene can be calculated from the gene pool. Comparing allele frequencies over time can help to determine if the population is evolving for that particular gene.

Genetic Equilibrium

If allele frequencies for a particular gene remain stable from generation to generation, then the population is described as being in a state of genetic equilibrium. Two statisticians working independently in the early 20th century, Godfrey H. Hardy and Wilhelm Weinberg, contributed to a mathematical equation, now known as the *Hardy-Weinberg equilibrium*, that describes the numerical relationship between the alleles. Hardy and Weinberg also contributed to a set of assumptions that were used to describe a population that was not evolving for a given gene:

1. A large population
2. No selection
3. No mutation
4. No migration
5. Random mating

Often, biology students wonder how any population can exist in Hardy-Weinberg equilibrium in nature, given the seemingly extreme conditions that must be met. It is essential to remember that the assumptions established by the Hardy-Weinberg equilibrium are applied only to the gene in question, not to all genes within the genome simultaneously.

Microevolution

When allele frequencies change over time, *microevolution* is occurring in that population. An allele frequency may increase due to a change in the environment that relates to an increase in the gene's evolutionary fitness; this breaks assumption 2 of the Hardy-Weinberg equilibrium. Or allele frequencies may change due to a disproportionate number of reproductively viable individuals that immigrate to the population (breaking assumption 4) or because of random events that are usually significant only in a very small population (breaking assumption 1).

Geographic Isolation

Whether it is in genetic equilibrium or not, a population may experience a major environmental change resulting in the population's being split into two smaller groups. If this geographic isolation is maintained for a long enough period of time, and if the now different environments experienced by the two populations place different pressures on the organisms, the populations can become genetically distinct as they accumulate adaptations and experience changes in allele frequencies.

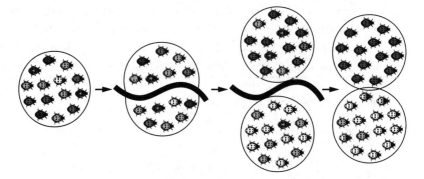

Allopatric speciation

 ## Speciation

When a significant number of genetic differences accumulate, one or both of the populations may undergo *speciation*, in which a new species emerges that is no longer able to produce viable offspring with a member of the parental species. Speciation is considered *macroevolution*, evolutionary change from the largest perspective.

The type of speciation that results from geographic isolation is known as *allopatric speciation* and is the most common form observed. Plants have an uncanny ability to participate in a second type of speciation described as *sympatric*. Here, no physical barrier separates individuals; instead, an abrupt change in chromosome number during meiosis can create a new species. Animals are typically unable to tolerate such drastic genetic shifts.

 It is helpful to consider the roots in the terms *microevolution* and *macroevolution*. Recall that *micro* means "small" and *macro* means "large," so *microevolution* refers to genetic change from the smallest perspective possible—that of an individual gene and the frequencies of the possible alleles. *Macroevolution*, then, describes evolution from the largest perspective—that of a species that is characterized by a certain set of genes composing its genome.

Exercise 10.3

If you can successfully work through the following questions, you should feel confident about being able to move on to the next section.

1. Which of the following could contribute to microevolution in a population?
   a. Natural selection
   b. A large population
   c. Random mating
   d. A lack of migration

2. Which of the following must be present to contribute to allopatric speciation?
   a. Mutation
   b. Interbreeding between populations
   c. Extinction
   d. A geographic barrier

3. Use the theme of the process and application of science to describe the concept of determination of allele frequencies in a gene pool.

   _____

   _____

   _____

11

Classification of Life

☑ Life has existed on Earth in one form or another for at least 3.5 billion years.

☑ Ongoing divergence among related species has resulted in a biosphere that supports more than seven million known species and likely at least four million to five million more unknown species!

☑ Taxonomy, the science of classification, is important in attempting to organize this vast quantity of species. Taxonomists are constantly updating classifications to keep up with new research and recently uncovered evolutionary evidence.

Eight Levels of Taxonomy

Taxonomy, the systematic method of naming and categorizing species, has likely been going on in one form or another for most of human history, as it is human nature to classify large volumes of information into meaningful subgroups. Scientifically speaking, however, taxonomy has existed since the 1700s, when the Swedish botanist Carolus Linnaeus devised the levels of taxonomy and system of binomial nomenclature still in use today.

Domain

Domain is the most inclusive of all the levels of classification. This *taxon* (group) was added well after Linnaeus established binomial nomenclature as a result of recent molecular evidence. Humans are in the domain Eukarya, as are other eukaryotic organisms. Prokaryotes are divided into two domains, Archaea and Bacteria.

Kingdom

Kingdom is the next most inclusive taxon (it was the most inclusive under the previous Linnaean system). Six kingdoms of life are currently recognized, an increase from the five-kingdom system of the recent past. Humans are in kingdom Animalia, as are all animals.

Phylum

A level down from kingdom is *phylum*. Humans are part of phylum Chordata (and subphylum Vertebrata). The Chordata phylum is so named because of distinct tissue that develops into the human spinal cord and spinal column.

Class

Just less inclusive than phylum is *class*. Humans belong to class Mammalia due to the unique mammary glands possessed by mammalian females that allow them to produce nurturing milk for their infant offspring.

Order and Family

More specific still, the next levels are *order* and *family*. Humans belong to the order Primates and the family Hominidae. Primates are classified by the

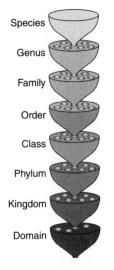

The levels of biological classification

presence of an opposable thumb for grasping and an increased reliance on vision with forward-facing eyes; hominids are bipedal, humanlike primates.

Genus and Species

The least inclusive levels, *genus* and *species*, are those required to identify a unique organism scientifically. These are the two levels that Linnaeus used in his system of binomial nomenclature. The genus name is used as the first part of the classification and is always capitalized. The species is the second name and is written in lowercase. Both names are written together and italicized, as in *Homo sapiens*.

The concepts of inclusivity and specificity can be puzzling when you try to consider and compare various taxonomic levels. It is easiest to start with the species taxon, as it is obviously the most specific of the eight levels and thus is the least inclusive. Conversely, the domain taxon is the most inclusive and the least specific.

Exercise 11.1

If you can successfully work through the following questions, you should feel confident about being able to move on to the next section.

1. Which of the following did Linnaeus *not* contribute to the science of classification?
   a. Binomial nomenclature
   b. Levels of taxonomy
   c. The domain taxon
   d. A way to systematically name species

2. The complete scientific name for a domestic cat would be _____.
   a. *Felis Catus*
   b. Felis catus
   c. *felis catus*
   d. *Felis catus*

3. Use the theme of the process and application of science to describe the significance of taxonomy.

   _____

   _____

   _____

Six Kingdoms of Life

As already explained, at the core of taxonomy is understanding relatedness between species. All information known about a species and its evolutionary history is used to group it appropriately with its close relatives. This approach establishes six recognized kingdoms of life, namely Archaebacteria, Eubacteria, Protista, Plantae, Fungi, and Animalia.

Archaebacteria

Members of this kingdom are exclusively asexual, unicellular prokaryotes. The extant species are restricted to extreme habitats considered uninhabitable by most life-forms and are thus reminiscent of the first bacteria to have evolved on Earth. Many Archaebacteria are heterotrophic, while some are uniquely chemoautotrophs, using the energy from simple, inorganic molecules to synthesize their own food.

Eubacteria

Members of this kingdom are exclusively asexual, unicellular prokaryotes too. So what makes them worthy of a separate kingdom from Archaebacteria? Almost everything else! Eubacteria lack *introns* (interrupting sequences within a gene that must be edited out) that members of Archaea and Eukarya possess, have uniquely structured rRNA and ribosomes, and are characterized by some degree of peptidoglycan in their cell wall. Eubacteria have diverse metabolic styles, including many heterotrophs (some parasitic and others decomposers), some photosynthetic photoautotrophs, and some that can switch between autotrophic and heterotrophic lifestyles. (Because Archaebacteria and Eubacteria possess similar superficial features, they were once classified together in the same—now nonexistent—kingdom of Monera.)

Protista

Members of this kingdom are diverse in body structure, metabolism, and lifestyle, so they are tough to classify. In fact, this kingdom is likely the next in line for revision, as protists are primarily classified by exclusion and actually represent a *polyphyletic* taxon (that is, one that does not exist naturally as its own branch in evolutionary history). Protists' diverse metabolic styles include some members that are plantlike and are thus photoautotrophs; others that are animal-like and eat actively; some that are parasitic; and yet others that are funguslike, eating dead and decaying matter.

Plantae

Members of this kingdom are strictly multicellular eukaryotes with true tissues making up their bodies and cellulose in their cell walls. The presence of true tissues distinguishes plants from their plantlike protist relatives; algae have specialized cells, but they do not organize into distinct structural regions of the protist as plant tissues do. Plants are characterized as photoautotrophs, capturing the sun's radiant energy and converting it into chemical energy for food. They thus provide the basis for many food chains.

Fungi

Members of this kingdom are exclusively heterotrophic eukaryotes with cell walls composed of *chitin*, a polysaccharide analogous to the cellulose present in plant cell walls. Most are multicellular, like the typical mushroom, truffle,

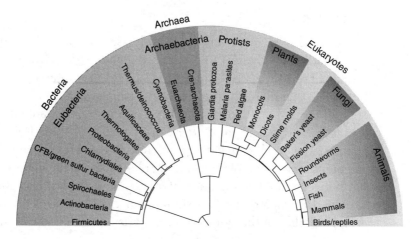

Summary of the phylogenetic tree of life

or lichen, while other notable varieties are unicellular, like the *Penicillium* mold and yeast. Most fungi have the ability to reproduce both sexually and asexually, depending on environmental conditions. They are *saprophytes*, major decomposers in most ecosystems.

Animalia

Members of this kingdom are exclusively multicellular, eukaryotic heterotrophs. Most animals reproduce sexually and have some (if not an extensive) degree of specialization of tissues. Animals are typically motile, although not always at all stages of life and not necessarily as adults. Helping this motility is the fact that their cells lack cell walls, so coordination between the cells can be very fast.

In trying to keep track of the various kingdoms of life and their associated characteristics, it is helpful to think in terms of evolution. Remember that Archaebacteria are the oldest, but Eubacteria diverged very early to establish their own kingdom. From the Archaebacteria base group diverged all members of the Eukarya domain, including Protista, Fungi, Plantae, and Animalia.

Exercise 11.2

If you can successfully work through the following questions, you should feel confident about being able to move on to the next section.

1. Which of the following characteristics is shared between typical members of the Fungi and Animalia kingdoms?
   a. Unicellularity
   b. Asexual reproduction
   c. Prokaryotic cells
   d. Heterotrophy

2. An individual plant and an individual protist might both be _____.
   a. photosynthetic
   b. unicellular
   c. autotrophic
   d. both a and c

3. Use the theme of the relationship between structure and function to describe the significance of the absence of cell walls in animals.

   _____

   _____

   _____

12

Microbes:
Viruses and Bacteria

 ☑ Viruses are considered strictly nonliving, as they do not meet all the characteristics of life. They do behave like organisms in many ways, and infections they cause can be deadly to life.

☑ Two of the six modern kingdoms of life are composed entirely of bacteria, namely Archaebacteria and Eubacteria.

☑ The members of Archaebacteria are most representative of the earliest cells to evolve. Today, they remain primarily in environments that most organisms consider too harsh for survival.

☑ Members of the Eubacteria kingdom are considered more modern than those of Archaebacteria. They diverged from the earliest cells and have taken up residence in nearly every habitat on Earth.

Viruses as Nonlife

Recall the six characteristics of life described in detail in Chapter 1. While viruses possess at least one characteristic of life (reproduction, if loosely defined), they fall short in meeting the remaining criteria used to establish something as living.

Viral Structure

The first key to understanding why viruses are nonliving is to consider their fundamental structure. Recall that all organisms must be composed of at least one cell; viruses are *acellular* (not composed of cells) and generally much simpler in structure than their cellular counterparts. Viruses have a protein coat called a *capsid* that protects the nucleic acid within. Some viruses also have an external membranous envelope composed of lipids from the last cell they infected.

Viral Replication

Instead of truly reproducing itself, a virus actually hijacks a host cell and reprograms it genetically to make and release more viruses. For that reason, it is often said to replicate, as the process is more similar to DNA replication than to reproduction.

Viruses that infect prokaryotes, called *bacteriophages*, use two different pathways to initiate replication in the host. The *lytic cycle* is immediately virulent (infectious): the virus simply lands on the bacterial cell and inserts its DNA into the host; the cell unintentionally begins to do replication and protein synthesis on the viral DNA to produce viral DNA and proteins. For viruses that instead contain RNA as their nucleic acid, the injected RNA molecule can immediately be used to make viral proteins through translation. The new viruses are assembled, and the cell eventually bursts and releases the pathogens.

Sometimes, the *lysogenic cycle* is employed first; this is referred to as a *latent period*, because no virus is actually produced and assembled. Instead, the viral DNA is incorporated into the host cell's main chromosome, where it is now called *prophage* while in this form. When this infected bacterium reproduces, it copies the entire chromosome—including the prophage—and passes it on to the offspring cells. A colony of bacteria full of individuals with prophage DNA can be produced very quickly. When conditions are appropriate, the prophage is cut out of the bacterial chromosome and enters the lytic cycle to actually make active viruses.

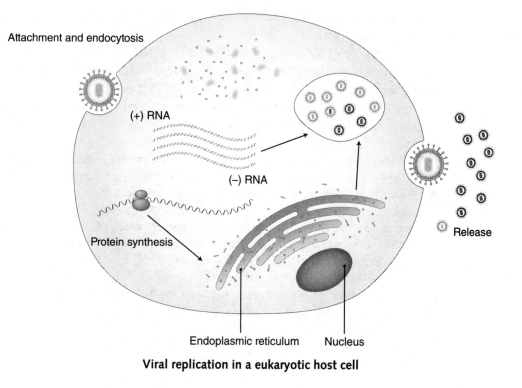

Attachment and endocytosis

(+) RNA

(−) RNA

Protein synthesis

Release

Endoplasmic reticulum   Nucleus

Viral replication in a eukaryotic host cell

Eukaryotic viruses use slightly different mechanisms to infect their specific hosts. Most animal viruses trick the host cell into engulfing them because of cell-surface receptors on the viral envelope. Once inside, the capsid is broken down, and the DNA is incorporated into the host cell's genome. From there, transcription and translation can produce more viral proteins, and replication can produce more viral nucleic acid. Once many copies of the virus have been assembled by the infected cell, the cell releases the virus in one of two ways: the viral load can increase to such a point that the cell bursts and viruses explode from the cell (as occurs during the lytic cycle in an infection of a prokaryote), or the cell can slowly release the virus by coating it with part of the cell membrane, creating the envelope for the next virus (similar to exocytosis of cellular proteins).

Viruses and Energy

An individual virus outside its host has no metabolism to speak of. It cannot actively move to find the next host. How does it then provide the energy to assemble new viruses during active infection? The answer is simple: it uses the energy supply of the host cell.

Because the virus lacks metabolism, it also cannot maintain homeostasis, grow, or respond to stimuli. It is essentially dormant when outside a host, but once it infects a host cell, that cell provides many of the initial benefits of life to the invading particle. Because a virus is truly dependent on its host, it is considered an *obligate intracellular parasite*.

 It is often assumed that viruses must have DNA as their nucleic acid to store genes. That is not the case, as some, in fact, contain RNA. These RNA viruses are tricky, because their single-stranded nucleic acid mutates much more quickly than the stable, double-stranded DNA molecule. This means the virus can overcome the human immune system, which often fails to recognize the virus after mutation. That actually accounts for why a new influenza vaccine is needed each year for ongoing protection from the flu!

 To keep track of the difference between the lytic and lysogenic cycles, it helps to consider the roots of the terms. To *lyse* something is to break it, as the explosion of viruses do to the cell after lytic infection. The term *lysogenic* implies not to break directly but instead to generate something that can eventually become lytic.

Exercise 12.1

If you can successfully work through the following questions, you should feel confident about being able to move on to the next section.

1. A typical virus might have all of the following *except* _____.
   a. a capsid
   b. a nucleus
   c. an envelope
   d. nucleic acid

2. Which of the following is true regarding viruses?
   a. Viruses are the simplest organisms.
   b. Viruses lack any structural organization, so they are considered nonliving.
   c. Viruses lack any ability to metabolize or grow, so they are considered nonliving.
   d. Viruses are descendants of the earliest bacterial cells.

3. Use the theme of continuity and variation to describe the mechanism of bacteriophage infection.

_____

_____

_____

Bacterial Biodiversity

Recall that prokaryotes were the lone inhabitants of the Earth for at least 2 billion years. This allowed plenty of time for divergence into many different lineages. One notable and major split occurred, establishing two major groups, the most ancestral bacteria known as Archaebacteria and the modern bacteria known as Eubacteria.

Archaebacteria: Thermoacidophiles

Thermoacidophiles prefer a habitat in which extremely hot temperatures (such as 100°C and above) and low pH are the norm. They inhabit volcanoes and hot springs, and they even provide the basis of deep-sea food chains around hydrothermal vents.

Archaebacteria: Halophiles

Halophiles thrive under high-salt conditions. The Dead Sea and Great Salt Lake are perfect habitats for such extreme prokaryotes, which have evolved adaptations that prevent dehydration.

Archaebacteria: Methanogens

Methanogens are Archaebacteria that inhabit anaerobic environments. As the group name implies, these organisms generate methane as a result of the metabolism of carbon dioxide (CO_2) and hydrogen gas (H_2). Restricted to anaerobic surroundings, methanogens inhabit places like swamps and the guts of cows and termites.

Eubacteria: Proteobacteria

Proteobacteria are a large and diverse group of Eubacteria. Many are animal pathogens, like *Salmonella* (which causes a form of food poisoning) and *Helicobacter* (which causes ulcers), while others are free-living. Some interesting proteobacteria are symbiotic with host cells; these include *enteric* (intestinal) *E. coli* that secrete vitamin K in the colon. Nitrogen-fixing *Rhizobium* bacteria actually carve out spaces within plant roots and there convert atmospheric nitrogen gas (N_2) into forms usable by fungi, plants, and eventually other organisms.

Eubacteria: Cyanobacteria

Ancestors of the modern *cyanobacteria* were responsible for creating and maintaining the first oxygen-rich atmosphere on Earth. This is the main group of photosynthetic bacteria; the blue-green pigment called *phycocyanin* characterizes the bacteria.

Eubacteria: Gram-Positive Bacteria

Like proteobacteria, gram-positive bacteria are a large and diverse group. Some species are pathogenic, like the *Bacillus anthracis* (which causes anthrax and can be used as a bioterrorist agent) and *Mycobacterium tuberculosis* (which causes tuberculosis). Others, like *Lactobacillus* species, carry out lactic acid metabolism and are used in the production of certain dairy products such as buttermilk and yogurt.

Practically speaking, Eubacteria can also be classified in many ways that are not necessarily phylogenetic (based on evolutionary relationships); this

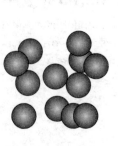

Bacilli   Cocci   Spirilli

Bacterial morphologies

can be helpful when a physician is attempting to diagnose a bacterial infection, for example. Such classifications include using shape and a Gram stain. Gram-positive bacteria, like the types already discussed, stain dark when the dye is applied, because they have thick cell wall material and thus appear purple or blue under the microscope. Gram-negative species, on the other hand, have a light stain due to a thin layer of cell wall sandwiched between two membranes.

The shape of the bacterial cell can also be useful in identifying species. A *bacillus* is rod-shaped, a *spirellum* spiraled, and a *coccus* is spherical. When many cocci are arranged in a chain, *streptococci* results. When they are arranged in a cluster, they are called *staphylococci*.

People often think they can take an antibiotic to help get rid of a viral infection. In fact, nothing is further from the truth. Antibiotics are designed to work against living bacterial cells, not viruses. Misusing antibiotics in this way when no bacterial infection is present is potentially harmful, because it contributes to bacterial resistance to the drug.

Exercise 12.2

If you can successfully work through the following questions, you should feel confident about being able to move on to the next section.

1. Which of the following is *not* a characteristic of any species of Archaebacteria?
   a. Loves heat
   b. Produces methane
   c. Is aerobic
   d. Loves salt

2. Some species of Eubacteria participate in all of the following roles *except*
   _____.
   a. nitrogen fixation
   b. enteric symbiosis
   c. deep-sea vent production
   d. photosynthesis

3. Use the theme of interdependence in nature to describe the evolutionary significance of cyanobacteria.

   _____

   _____

   _____

13

Protists and Fungi

☑ All protists and fungi are composed of eukaryotic cells and are thus part of the domain Eukarya.

☑ Organisms are placed in the kingdom Protista based on exclusion from other kingdoms, and subgroups of protists are often based more on ecological roles and less on relatedness. These nonphylogenetic means of grouping protists leave this kingdom as the next up for revision by taxonomists.

☑ Organisms are placed in the kingdom Fungi based on the following set of characteristics: absorptive heterotrophy, a cell wall made of chitin, and excess sugar storage in the form of glycogen.

☑ Most fungi, such as mushrooms and truffles, are multicellular, but notably unicellular varieties exist and are abundant. They include molds and yeasts.

Protist Biodiversity

The protist kingdom displays the highest degree of biodiversity of any kingdom; some of this can be attributed to the way in which protists are classified. Organisms are often classified as Protista based on their exclusion from other kingdoms and not because they are truly phylogenetic (closely related through evolution). For this reason, Protista is currently under revision by taxonomists.

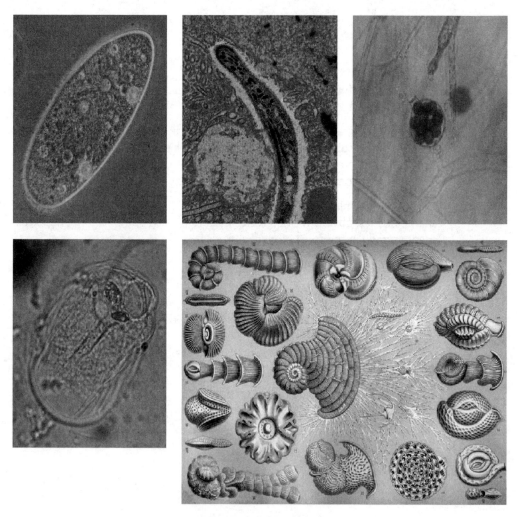

Protist biodiversity

The Evolution of Early Protists Through Endosymbiosis

For at least 2 billion years, bacteria ruled the planet. At some point, a phagocytic bacterium consumed an aerobic bacterium, likely intending to break it down for energy. It was unable to do so, and the aerobic bacterium continued to function normally inside the larger cell. Over time, the two bacteria became interdependent to the extent that they emerged as a new cell form—eukaryotic. The internal bacterium became the mitochondrion.

This process of endosymbiosis also occurred later with a photosynthetic cyanobacterium; this time, the internal bacterium became the chloroplast.

Animal-Like Protists

Protists in this group are like animals in that they are predominantly active heterotrophs. Animal-like protists differ from animals, however, in that they are exclusively unicellular. Some animal-like protists, like amoebas, move by forcing cytoplasm into one end of the cell and reshaping the cell membrane around it to create a *pseudopod*, or false foot. The pseudopodia can also be used to engulf food particles in a phagocytic mechanism.

Ciliates, including the well-studied *Paramecium*, have *cilia*, short, hair-like projections that surround the entire cell surface. They use cilia to scoot around in an aqueous environment and to direct food into their oral groove for endocytosis into a food vacuole. Tail-like flagella propel yet other types of animal-like protists around their aqueous environments; for example, this is true of *Giardia*, which causes intestinal disorders in humans who drink infested waters.

Some animal-like protists have lost the ability (and need) to move because they have adapted to become parasites; the well-known and highly feared *Plasmodium* that causes malaria is one such protist.

Plantlike Protists

Plantlike protists are exclusively photoautotrophs and are commonly referred to as *algae*. However, algae display much more diversity in photosynthetic pigments than do true plants. Algae range in color from red to green to brown, and they also range in body type from unicellular to colonial to multicellular (including seaweeds and kelp). They inhabit marine and freshwater ecosystems as well as soil and other places where water is abundant.

Funguslike Protists

Funguslike protists comprise two main varieties: slime molds and water molds. Slime molds are unicellular, absorptive heterotrophs that primarily serve the ecological role of decomposers. Water molds are unicellular or filamentous, and all are parasitic.

Because of the way protists are traditionally classified, it is easy to misunderstand their evolutionary relationships. For taxonomists, it is important to classify according to phylogenetics. This means fundamentally regrouping protists such that the animal-like, plantlike, and funguslike divisions will no longer be represented. It is still useful, however, to describe them this way when referring to their ecological roles.

Exercise 13.1

If you can successfully work through the following questions, you should feel confident about being able to move on to the next section.

1. All of the following are means of locomotion in animal-like protists *except* _____.
   a. cilia
   b. pseudopodia
   c. flagella
   d. centrioles

2. Which of the following types of protists is (are) most likely to be misclassified as a plant?
   a. Slime mold
   b. Seaweed
   c. *Paramecium*
   d. Colonial algae

3. Use the theme of the process and application of science to describe the concept of reclassifying protists.

   _____

   _____

   _____

Fungus Biodiversity

Fungi are everywhere, but they are often overlooked. Some, like yeasts and molds, are unicellular and only become apparent after reproduction when a mass appears. Even the large, multicellular fungi, like mushrooms, are recognizable to humans only when they peek out aboveground to release reproductive spores. One of the largest known organisms on Earth is a honey mushroom in Oregon; although it is roughly 3.5 square miles in area, the vast majority of the organism is composed of underground *mycelia*. Each mycelium, in fact, is made up of masses of intertwined *hyphae*, elongated filaments that provide the fundamental basis of fungal structure.

Hyphae are described as either *septate*, possessing obvious cross walls between cells, or *coenocytic*, lacking such cross walls and thus appearing more fluid from cell to cell. To facilitate nourishing the fungus, a hypha releases digestive enzymes into the soil, right onto its food. After the food—

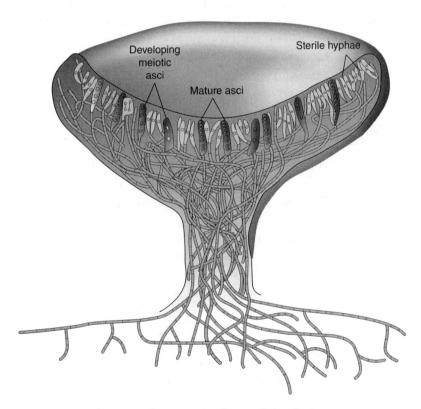

A sac fungus with mating hyphae and developing spores

usually some dead organic matter—is broken down, the monomer products are absorbed right into the hypha.

In addition to the feeding process, hyphae function in reproduction. Fungi usually possess hyphae of either a plus or minus mating type. Hyphae of opposite mating types must merge in order to sexually reproduce and increase genetic variability. A fungus actually spends most of its life cycle with haploid cells, but two haploid hyphae of opposite mating types can fuse to produce a *dikaryon* phase. This phase is not considered diploid, because the two haploid nuclei do not fuse for some time. Eventually, a true diploid structure is produced and then meiosis occurs to produce sexual spores.

Chytridiomycota

Fungi in this group, commonly referred to as *chytrids*, are likely representative of the earliest fungi to evolve, and they used to be classified as protists. Chytrids reside in aqueous habitats—some are freshwater, others are marine. All chytrids use flagellated gametes to facilitate fertilization in an aqueous environment. Many are typically unicellular, while others are filamentous and produce a coenocytic hypha, as do other fungi. Many members of this phylum are saprophytic, meaning they decompose dead and decaying organic matter, while others—including the potato wart—are plant or alga parasites.

Glomeromycota

Glomeromycota fungi are composed of coenocytic hyphae. Many fungi in this phylum participate in an essential symbiotic relationship with cortical root cells of vascular plants called *arbuscular mycorrhizae*. The filaments of the glomeromycetes extend into and around the root cells—or potentially even around the thallus of a bryophyte (a simple plant, like a moss). By doing so, they vastly increase the surface area of the roots and thus facilitate increased absorption of water and minerals from the soil. At least 200 different species of glomeromycetes are widely distributed throughout soils worldwide; up to 80 percent of land plants may have symbiotic relationships with these fungi.

Other fungi in this phylum include those that were formerly classified as zygomycota, an old phylum no longer in use. These include the black bread mold and many other microorganisms that decompose organic matter.

Shelf fungi, a common type of club fungus

Basidiomycota

Fungi in this phylum are commonly characterized as *club fungi* because of their club-shaped basidia that produce sexual spores through meiosis. The fungi are composed of coenocytic hyphae and are major ecosystem nutrient recyclers. Mushrooms and shelf fungi are examples of organisms classified into this phylum.

Ascomycota

Fungi in this group are commonly called *sac fungi* because of their distinct, sac-shaped reproductive structure. They possess septate hyphae if multi-cellular or an elongated single hypha if unicellular. This phylum contains the most species of any group of fungi, including many that parasitize humans and other animals (such as those fungi that cause athlete's foot or ringworm). Other members of this phylum, like truffles and morels, are consumed by humans.

Although not unique to this phylum, most lichens are Ascomycota. A *lichen* is a composite organism that represents a mutually beneficial symbi-otic relationship between a fungus and a unicellular photoautotroph. This photosynthetic partner is usually a form of green algae or cyanobacteria, contributing monosaccharides to the fungus for energy and gaining a protective, stable environment in return.

Deuteromycota

Because fungi are traditionally classified according to their mode of sexual reproduction, those that only reproduce asexually do not fit in neatly. They are instead placed in a nonphylogenetic group called Dueteromycota, or *imperfect fungi*. They are typical fungi in every other way, so the common name is a bit unfortunate.

It is helpful to know that some of the names of the fungal phyla are derived from the name of their highly specialized sexual reproductive structures. For example, the Basidiomycota possess basidia that generate sexual basidiospores through meiosis. The Ascomycota, on the other hand, produce ascospores in their asci, but the fundamental processes and life cycle are the same.

Exercise 13.2

If you can successfully work through the following questions, you should feel confident about being able to move on to the next section.

1. Which of the following is *not* true of a fungal hypha?
   a. It can be coenocytic.
   b. It can be septate.
   c. It is usually diploid.
   d. It is involved in digestion and absorption of nutrients.

2. Each of the following involves Ascomycota *except* _____.
   a. a lichen
   b. a mushroom
   c. a sac fungus
   d. athlete's foot

3. Use the theme of interdependence in nature to describe the ecological significance of fungi symbioses like mycorrhizae and that of lichens.

   _____

   _____

   _____

14

Plants and Animals

☑ Plants and animals are among the largest and most complex organisms on Earth. They are strictly eukaryotic and thus members of the domain Eukarya.

☑ Plants are characterized as multicellular photoautotrophs.

☑ True plants can be distinguished from plantlike protists by the presence of true tissues. Although algae can have specialized cells, they are not typically aggregated into a functional tissue.

☑ Animals are characterized as multicellular, ingestive heterotrophs.

☑ A true animal can be distinguished from an animal-like protist by the presence of a multicellular body plan. Protozoans are typically unicellular.

Plant Biodiversity

Plants only fully reproduce after completing an *alternation of generations*, meaning their life cycle consists of one phase in which they are haploid and produce gametes (called the *gametophyte generation*) and another phase in which they are diploid and produce spores (called the *sporophyte generation*). Some plants are gametophyte-dominant, spending the majority of their lives in the haploid phase. Others are sporophyte-dominant, composed of diploid cells for the majority of their lives. Regardless, a plant must go through both phases in order to produce the next-generation plant.

Green Alga Ancestor

A green alga, and thus a member of the protist kingdom, is the likely ancestor of all plants. Green algae and plants share many characteristics. Both possess the same collection of photosynthetic pigments (chlorophyll *a*, chlorophyll *b*, and carotenoids); have cell walls composed of cellulose; and store excess sugar as starch.

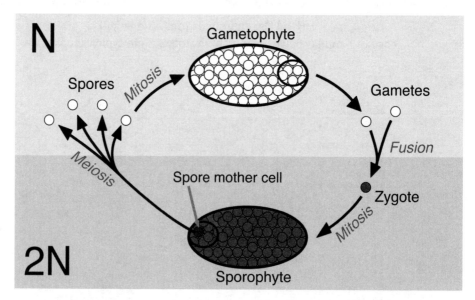

Summary of the alternation of generations. Key: N = the haploid (gametophyte) phase; 2N = the diploid (sporophyte) phase.

Bryophytes

Mosses and other bryophytes are the oldest existing plant group. Bryophytes include simple plants like mosses, liverworts, and hornworts. They are relatively small in stature, rarely exceeding 5 to 10 centimeters in length. This is because they lack vascular tissue, an internal network of tubes that provides support and circulation of internal fluids. A typical adult moss is made up of haploid cells; it is thus considered gametophyte-dominant.

Seedless Vascular Plants

Ferns and other seedless vascular plants evolved slightly later than early bryophytes. Seedless vascular plants grow much larger than bryophytes because the former possess vascular tissue. Some, like tree ferns, grow very large as they compete in thick rain-forest canopies. A typical adult fern is composed of diploid cells; it is thus considered sporophyte-dominant. Ferns are unique and recognizable by the clusters of spore-producing structures called *sori* that dot the underside of their leafy fronds. When mature, each sorus bursts open and releases haploid spores that disperse some distance away from the parent plant to help continue the life cycle.

Gymnosperms

Early conifers and other gymnosperms were the next plant group to evolve. *Gymnosperms* include some of the tallest and most massive organisms on Earth, the coastal redwoods and giant sequoias, respectively. Conifers often grow in massive single-species stands and were among the first to use wind-blown pollen as a means of transferring sperm between male (pollen) cones and the egg within the female (ovulate) cones. Adult gymnosperms are heavily sporophyte-dominant, so much so that the gametophyte phase is reduced to a few cells.

Angiosperms

Finally, *angiosperms* evolved thanks to the highly adaptive flower structure. Some are *monocots*, plants that possess a single seed leaf during early development, while others are *dicots*, plants that instead possess two seed leaves. Flowering plants include monocots like corns, grasses, and bamboos, as well as dicots such as roses and all varieties of fruit trees. The flower, acting as a

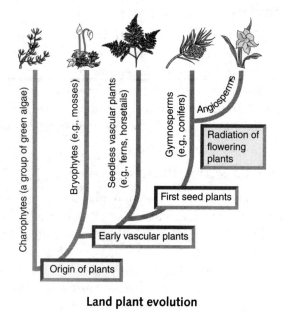

Land plant evolution

visual and olfactory attractant to animal pollinators, conferred such a reproductive advantage to its species that these plants quickly diverged and now
represent the living plant group with the greatest number of extant species. Like the gymnosperms, angiosperms are overwhelmingly sporophyte-
dominant and undergo a reduced gametophyte phase.

Plant biodiversity is best understood while also examining the evolutionary move from an aquatic existence to a terrestrial one. The green algae that
gave rise to plants were—and still are—well suited to a world in which water
surrounds them. For them to adapt effectively to much drier conditions and
ensure survival, many structures had to evolve. First, terrestrial plants
needed to be able to resist desiccation (drying out) of both the plant body
and the reproductive cells. The evolution of the cuticle met this demand;
this waxy secretion from the epidermis of plants contributes to a watertight
barrier.

Another problem arose from the presence of the cuticle, for how can gas
exchange take place across such a surface? Stomata and guard cells then
evolved; the *stomatal pores* allow gas exchange (and water loss) to occur
along the leaf, while the *guard cells* regulate each opening to prevent desiccation. Complex tissues in roots, stems, and leaves also evolved for specialized life functions in plants.

Exercise 14.1

If you can successfully work through the following questions, you should feel confident about being able to move on to the next section.

1. Which of the following plant groups and characteristics are mismatched?
   a. Angiosperms—fruits
   b. Seedless vascular plants—sori
   c. Gymnosperms—gametophyte-dominant
   d. Nonvascular plants—reduced size

2. All of the following structures are significant in the adaptation of plants to terrestrial life *except* _____.
   a. stomata
   b. vascular tissue
   c. the cuticle
   d. chloroplasts

3. Use the theme of interdependence in nature to describe the concept of angiosperm pollination.

   _____

   _____

   _____

Invertebrate Biodiversity

Animals exhibit a wide range of body types, lifestyles, and habitats. Many major phyla exist today; the most significant ones (evolutionarily, ecologically, and so on) are summarized here. The vast majority of animal phyla are *invertebrates*, meaning they lack a vertebral column to protect the nerve cord. Only phylum Chordata contains all *vertebrates*, or animals with this protective structure.

A Colonial Choanoflagellate Ancestor

Over time, many individual animal-like protists called *choanoflagellates* likely formed a colony within which individual organisms benefited from the functions of their partners. Such successful colonial existence eventually gave rise to all members of the kingdom Animalia as the individual cells became completely interdependent.

Sponges and Cnidarians

Evolutionarily, the oldest animal phyla include the sponges and the cnidarians. Sponges, usually all considered members of the phylum Porifera, are among the simplest animals and lack true tissues. They have specialized cells, like their characteristic feeding choanocytes, but these cells are scattered throughout the body and do not collect into conspicuous tissues. Sponges are unique in the animal kingdom in their possession of an asymmetrical body plan.

Cnidarians, such as corals, sea anemones, and sea jellies, are characterized by the presence of true tissues and are, in fact, considered *diploblasts* because they possess two distinct tissue layers early in embryonic development. This contributes to their characteristic radial symmetry and to the development of their saclike body plan with only one opening to the gut. Cnidarians are often recognized for their stinging capacity, because they possess stinging cnidocytes equipped with a spring-loaded trigger.

Both sponges and cnidarians are capable of sexual reproduction through the fertilization of eggs and sperm, and also of asexual reproduction through methods like budding and fragmentation.

Flatworms, Roundworms, and Segmented Worms

Flatworms are also diploblasts and have a saclike body plan, but unlike cnidarians, they have bilateral symmetry. A typical flatworm, a planarian, possesses a short external tube near its midline that acts as the intake point for nutrients as well as the elimination point for wastes. Given that there is no third embryonic tissue layer, flatworms cannot develop a body cavity and are considered *acoelomates*.

Roundworms are triploblasts, indicating their possession of three distinct embryonic tissue layers. They are also *pseudocoelomates*, meaning that their third embryonic tissue layer allows for development of a body cavity, or *coelom*, but the coelom does not develop from the mesoderm (middle) layer. The body cavity permits a tube-within-a-tube body plan with a complete digestive tract. A diverse and abundant group, roundworms include some parasites like pinworm and the worm responsible for trichinosis.

Segmented worms, like the common earthworm, are complex triploblastic animals with a true coelom. They possess a complete digestive tract with a high degree of specialization throughout. Segmented worms also have an efficient closed circulatory system with a set of small pumps called *aortic arches* to encourage blood flow. Many are hermaphroditic, possessing both male and female sex organs. During reproduction, each worm exchanges sperm with the other, thereby doubling their reproductive success.

Arthropods and Echinoderms

Arthropods and echinoderms are among the most complex invertebrates. *Arthropods* are the most diverse animal phyla, with members including insects, crustaceans, and arachnids. They are true coelomates with bilateral symmetry and highly specialized jointed appendages attached along the body. Arthropods inhabit almost every ecosystem on Earth.

Echinoderms are so named for their characteristic spiny skin—think of sea stars and sea urchins. Given their bottom-feeder lifestyle, adult echinoderms have radial symmetry. Due to a novel shared embryonic development pattern, echinoderms are the animal phylum most closely related to our own. These animals are strictly marine, so they have special adaptations for such an environment.

It is easy to confuse all the animal development patterns and strategies. The key here is to think evolutionarily. If an animal has true tissues, then it has either two or three tissue layers in the early developing embryo. Only if it is triploblastic can the animal also be a true coelomate.

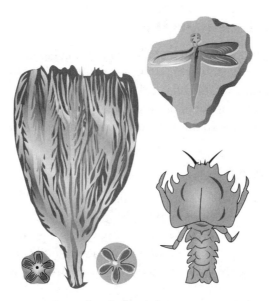

Invertebrate fossils from the Jurassic period

Exercise 14.2

If you can successfully work through the following questions, you should feel confident about being able to move on to the next section.

1. Which of the following phyla includes animals that are diploblastic with a saclike body plan?
   a. Flatworms
   b. Roundworms
   c. Cnidarians
   d. Sponges

2. If an animal is a coelomate, which of the following must also be true?
   a. The animal has true tissues.
   b. The animal is triploblastic.
   c. The animal is bilaterally symmetrical.
   d. Both a and b are true.

3. Use the theme of continuity and variation to describe the trends of symmetry observed across animal invertebrate phyla.

   _____

   _____

   _____

Vertebrate Biodiversity

The phylum Chordata includes complex coelomate animals that are characterized by the following at some point in their life cycle: a dorsal, hollow nerve cord; a rod of tissue called the *notochord*; a postanal tail; and gill slits. The vast majority of chordates are also vertebrates, although a few groups of invertebrate chordates exist today. These include *lancelets* and *tunicates*.

Chordates

Within the chordate phylum, the vertebrate subgroup diverged extensively over time. In addition to possessing the typical chordate features, these animals are characterized by protective coverings around the spinal cord (vertebrae) and brain (cranium). Vertebrates usually have many specialized body systems that are best adapted to their environment or habitat.

Cartilaginous and Bony Fish

Both cartilaginous and bony fish are some of the oldest extant vertebrate groups. Cartilaginous fish include sharks, skates, and rays. They do possess an endoskeleton, but it is made of cartilage, not mineralized bone. As far as reproduction and development go, cartilaginous fish are usually associated with internal fertilization and *ovovivipary*, the ability of a female to give live birth to her offspring after they hatch from eggs internally.

As their name implies, bony fish possess a fully mineralized endoskeleton. They also possess a swim bladder that is lacking in their cartilaginous counterparts (which rely instead on an extra-fatty liver to float). This structure contains internal gases and allows a bony fish to maintain buoyancy more readily. Unlike sharks and rays, bony fish are characterized by external fertilization in a mass spawning event.

Both groups of fishes possess a *lateral line*, a means of detecting vibrations in the water and responding appropriately (the appropriate response will depend on whether an organism is a predator or a prey species). Only cartilaginous fish, however, possess *ampullae of Lorenzini*, specialized electroreceptors that allow them to detect electrical impulses given off by the heartbeat and muscular activity of nearby prey.

Both groups are also characterized as *ectothermic*, meaning their body temperature is determined by the external environment. The possession of a two-chambered heart means that the circulatory system of both cartilaginous and bony fish is composed of a single loop that distributes oxygen to the body's tissues after picking up a fresh supply of oxygen in the lungs.

Amphibians and Reptiles

Amphibians and reptiles were the next groups of vertebrates to evolve. *Amphibians* represent the first major group of *tetrapods*, animals with paired legs instead of fins. While this indicates a terrestrial lifestyle, amphibians are still strongly tied to an aqueous habitat as well. Consider the classic case of the frog and its complex metamorphosis: the first phase of its life cycle is spent as a tadpole in freshwater, while the adult phase consists of an animal capable of moving between water and land.

Amphibians, like fish, are characterized by *ectothermy*, but they possess a three-chambered heart instead of a two-chambered one. This means that amphibians have two loops in the circulatory system; one is dedicated to picking up fresh oxygen from the lungs (or external gills) and skin and directing it to the heart, and the other is dedicated to delivering the oxygenated blood to the body's cells for respiration.

Reptiles represent the first group of vertebrates to adapt fully to terrestrial life. Their significant structures include the waxy, *keratinized* scales that provide a watertight outer surface and the amniotic egg that resists drying after being laid by the mother. Keratin is a protein that is tough and impermeable to water, so it makes for an effective barrier. Reptiles are ectothermic and demonstrate variation in heart structure. Most have three chambers with a partial septum separating the lower ventricle, but the crocodilians have evolved a complete septum and thus a four-chambered heart. The more complete the septum, the higher the degree of separation of oxygen-rich and oxygen-poor blood, and so the more efficient the organism is energetically.

Birds and Mammals

Finally, birds and mammals represent the extent of vertebrate biodiversity. *Birds* diverged directly from their reptile ancestor, a feathered, flying dinosaur! Birds have hollow bones and flow-through lungs to help with flight. Instead of scales, they evolved keratinized feathers.

Mammals are characterized by their keratinized covering of skin and hair, as well as the nourishing mammary glands for which they are named. Three major groups of mammals exist: the rare, egg-laying monotremes (such as echidnas and platypuses); the pouch-bearing marsupials (such as kangaroos and opossums); and the placental mammals (such as horses and gorillas).

Because of the high energy needs of both birds and mammals, both of these classes of vertebrates have highly specialized body systems for increased efficiency. They are also the only two vertebrate groups characterized by endothermy, meaning they can control their body temperature physiologically. Controlling body temperature is a major way of controlling metabolism and helping the coordination of organ systems, but it also requires more of an energy input from the diet. Both mammals and birds possess a four-chambered heart to maximize energy efficiency through the separation of oxygenated and deoxygenated blood. They also invest the highest degree of parental care in their offspring of any vertebrate groups.

Vertebrate biodiversity

Many characteristics are shared by several animal phyla, but exceptions to the rule typically exist. These exceptions are often unique and resourceful adaptations that have evolved to fit a specific niche in a certain ecosystem.

 # Exercise 14.3

If you can successfully work through the following questions, you should feel confident about being able to move on to the next section.

1. Which of the following correctly pairs the vertebrate group to the number of chambers in the heart?
   a. Mammals—three
   b. Reptiles—three or four
   c. Bony fish—three
   d. Birds—three or four

2. All of the following vertebrates are ectothermic *except* _____.
   a. reptiles
   b. birds
   c. cartilaginous fish
   d. amphibians

3. Use the theme of regulation: feedback and response to describe the concept of endothermy.

   _____

   _____

   _____

15

Plant Structure and Function

☑ Plants are multicellular, eukaryotic photoautotrophs.

☑ Plants evolved from a green algal ancestor that also possessed the same collection of photosynthetic pigments, stored excess sugar as starch, and had cell walls composed of cellulose.

☑ Plant cells can be distinguished from animal cells by their possession of *plastids* (including the chloroplast); a large, central vacuole; and a cell wall.

☑ Plants are a composite of two organ systems, three basic organs, three tissue types, and three generalized cell types.

☑ Plants are much more complex and active than they may seem. They can grow in interesting ways and can rearrange their body parts to optimize their metabolism. They also have essential and sophisticated interactions with fungi and animals.

Plant Physiology

Plants have a high degree of organization, especially when compared with their algal ancestors that lacked true tissues. Here, we consider the plant body from the perspective of decreasing levels of biological organization.

An Individual Plant

An individual plant constitutes an organism. Any member of the plant kingdom falls into one of four major groups: bryophytes (such as mosses); seedless vascular plants (such as ferns); gymnosperms (such as conifers); and angiosperms (such as lilies).

The Root and Shoot Systems

An individual plant is generally composed of two organ systems: the root system and the shoot system. The *root system* includes any of the plant organs that lie belowground; thus, it is generally focused on absorption of materials from the soil. The *shoot system*, on the other hand, includes any plant organs that lie aboveground. Consequently, these structures focus primarily on photosynthesis and reproduction.

Roots, Stems, and Leaves

Roots, stems, and leaves constitute plants' three basic organ types; roots make up the majority of the root system, and the stem and leaves make up the shoot system. Leaves are typically broad and thin to maximize their efficiency at trapping light energy, much like a solar panel. Stems are elongated and designed to prop up the leaves closer to the light source. Roots are highly branched, with increased surface area for the absorption of water and dissolved minerals.

Dermal, Vascular, and Ground Tissues

Plants use combinations of dermal, vascular, and ground tissues to make up their organs. Dermal tissue is found along the external lining of the plant body and consists mainly of epidermal cells and guard cells; its functions are protection and regulation of water loss. Vascular tissue makes up the *xylem* and *phloem* tubes that function in circulating materials around the plant body. The rest of the plant comprises ground tissue. Photosynthesis, storage of materials, and support of the plant body are the major functions of ground tissue.

Parenchyma, Collenchyma, and Sclerenchyma Cells

Plant tissues are composed of one or more of three general cell types: parenchyma, collenchyma, and sclerenchyma. *Parenchyma* cells have a standard primary cell wall. They are alive at functional maturity and make important contributions to the plant's metabolic activities. *Collenchyma* cells, also alive at functional maturity, possess a thickened primary cell wall and a standard secondary cell wall; they provide flexible support of the plant body. *Sclerenchyma* cells are dead at functional maturity and possess thick, irregular secondary cell walls. They function primarily to support the plant.

Each of the general cell types can specialize in structure and function to produce a wide variety of differentiated cells.

The generalized cell types (such as parenchyma) are just that—generalized, not specialized. This means that the highly specialized, conductive vessel elements in the cell are differentiated from sclerenchyma cells, cells that are dead at functional maturity. No energy is needed to move water and dissolved solutes up the xylem tubes, so they can become hollow and focus on permitting capillarity and encouraging maximum flow.

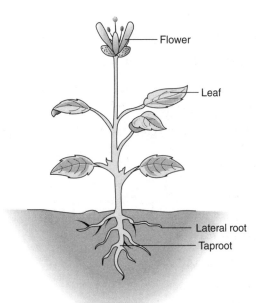

Basic plant anatomy

Roots are structured so they can best absorb water and minerals from the soil and transport the solution to the stem. Within the epidermal layer, tiny extensions of the cell wall (called *root hairs*) work to vastly increase surface area for absorption.

Roots generally occur in one of two types: fibrous and taproots. When a sprouting seed's primary root emerges to become the largest, central root, it is called a *taproot*. Other, smaller roots emerge from the taproot; these are known as *lateral roots*. When many individual roots emerge from the stem and grow parallel to the primary root, they are described as *fibrous*.

Stems generally constitute the major length of the plant body and are there to support the leaves toward the sun and to transport xylem sap to the leaves for photosynthesis. Some specialized stem types, such as the cactus stem, have evolved. This storage stem is also the primary photosynthetic organ, because the cactus leaves have become protective spines. *Stolons* are modified horizontal stems that extend out from the parent plant to create asexual offspring. These are readily observed in the strawberry plant.

Leaves are generally structured to be broad and planar so they can most efficiently trap sunlight for photosynthesis. Some leaves are simple and composed of one major lobe, whereas others are called *multiple leaves*. In any case, the stem attached to the leaf is called the *petiole*, and the rest is the *blade*.

Exercise 15.1

If you can successfully work through the following questions, you should feel confident about being able to move on to the next section.

1. All of the following are considered a plant tissue type *except* _____.
   a. vascular
   b. sclerenchyma
   c. ground
   d. dermal

2. Of the following plant structures, which is the most inclusive organizationally?
   a. Parenchyma
   b. Root
   c. Vascular tissue
   d. Shoot system

3. Use the theme of the relationship between structure and function to describe the concept of plant organs.

_____

_____

_____

Plant Characteristics

Plants often seem simpler from the outside than they actually are. Rooted in place, they appear inactive relative to their animal counterparts. In reality, plants use hormones to direct movement and growth in response to their environment on both a daily and a seasonal basis. Plants also are, in some rare cases, able to consume animals and parasitize fungi.

Organization: From Cells to Body Systems

Plants exhibit a high degree of organization, as already discussed. From cells to tissues, tissues to organs, organs to organ systems—all are essential components to a functional plant organism. From the simplest bryophytes to the most complex angiosperms, plants are exclusively multicellular, and all of them possess some shoot and root system that enables their lifestyle.

Metabolism: Photosynthesis

Almost without exception, plants are photoautotrophs. Of the nearly 300,000 named species, only a very few are either partially or exclusively parasitic. These include mistletoe, which retains the ability to make a small quantity of its own food, and the Indian pipe plant, which lacks any photosynthetic pigments and often appears white. The vast majority of plants obtain energy to build organic compounds from the sun using chlorophyll, so their green color is usually a dead giveaway.

Reproduction: Fruits and Flowers

The angiosperm flower is highly beneficial to the reproductive fitness of the plant. Its brightly colored petals attract specific pollinators, which then unknowingly pick up some pollen to transfer to another plant. This coordinated pollination gives angiosperms a clear evolutionary advantage over other plants.

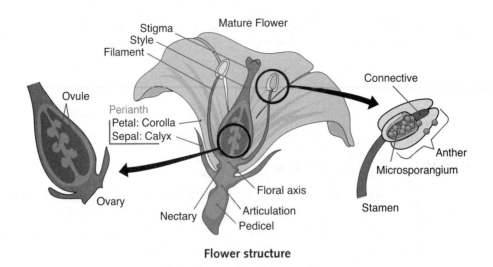

Flower structure

If the flower eventually develops into a fruit through the thickening and maturation of the ovary, then the seeds are packaged into a tasty and protective treat. A tempted animal ingests the fruit, swallows the seeds, and then eliminates them as waste after digestion. The tough seed coat protects the seed from the harsh conditions of the digestive tract, and this covering aids in dispersal of the seed. The young plant embryo can now germinate, grow, and eventually reproduce itself in a region potentially distant from the parent plant. This decreases the chance of inbreeding and contributes to genetic variation.

Growth: Meristems

Meristems are regions of active mitosis within the plant body. Because of a plant's specialized metabolic and nutritional needs, it is mainly concerned with growing up toward the sun and down toward water. This primary growth is due to the activity of apical meristems at the tips of the roots and shoots.

Only some *woody plants* (like trees) are capable of secondary, lateral growth. Other meristems, called *vascular cambium* and *cork cambium*, are responsible for creating new vascular tissue and cork (or bark), respectively. The vascular cambium creates new growth rings observable in tree trunks.

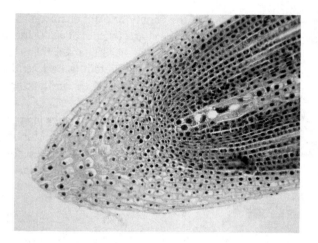

A photomicrograph of the root apical meristem in a plant

Homeostasis: Regulation of Transpiration

Transpiration is the movement of water from the roots of the plant to the leaves, where it is required for photosynthesis. The process is regulated by the guard cells on the underside of the leaves; these cells allow the stomata to be in either an open or a closed position. If they're open, water can evaporate from the air spaces inside the leaf to the external atmosphere. While plants don't want to lose too much water, they do need some to transpire out of the stoma so the column of water in the xylem is pulled into the leaf. This in turn encourages more water to flow into the roots.

Response to Stimuli: Tropisms

While plants can't generally relocate their entire body, they can encourage growth in particular ways so they can move toward or away from a stimulus, a movement referred to as a *tropism*. In general, if the movement is toward the stimulus, then it is characterized as positive; when instead the movement is away from the stimulus, it is called negative. When light strikes one side of a plant disproportionately, a hormone called *auxin* moves to the shady side of the stem. There, it acts by elongating cells on that side. The extra slack that accumulates causes the plant stem to bend toward the light—positive phototropism.

Other tropisms include *thigmotropism*, growth in response to a touch stimulus, and *gravitropism*, growth in response to the force of gravity.

If keeping track of the variety of plant characteristics given here seems difficult, just remember the characteristics of life. Plants have myriad interesting and diverse adaptations that have enabled them to survive in all sorts of environments and ecosystems. Referring back to the characteristics of life will help you generate examples of the more specific plant characteristics.

Plants like the Venus flytrap and the pitcher plant are considered carnivorous because they consume animals. It is important to realize that, although they do trap small animals like insects and digest them with enzymes, they evolved this feeding method as a means of obtaining minerals because their soil was deficient. They are still able to photosynthesize to process organic compounds.

Exercise 15.2

If you can successfully work through the following questions, you should feel confident about being able to move on to the next section.

1. Where might you expect to find a meristem?
   a. At the tip of the root
   b. Just under the bark of a tree trunk
   c. In the oldest part of the shoot
   d. In a mature fruit

2. If a plant moves away from a physical touch stimulus, this would constitute _____.
   a. positive phototropism
   b. negative thigmotropism
   c. positive thigmotropism
   d. negative phototropism

3. Use the theme of regulation: feedback and response to describe the action of guard cells.

   _____

   _____

   _____

16

Animals and the Organ Systems

☑ Most animals have highly organized structures and display a significant degree of specialization within their organ systems.

☑ Only sponges lack true tissues altogether and demonstrate specialization only on the cellular level; cnidarians lack true organs but have specialized tissues.

☑ Most other animals possess at least some organs and organ systems for increased efficiency of function. Up to 11 different organ systems are observed in animals.

Animal Organization

As with all organisms, animal structure is characterized by the increasing levels of organization present. All animals are multicellular, but not all actually possess true tissues. Most animals, however, have not only tissues but also organs arranged into organ systems that together contribute to overall functioning of the individual organism.

An Individual Animal

Animals are organized into many different phyla, 13 of which are typically considered major. These organisms include sponges, segmented worms, arthropods, and chordates. Animals are classified based on their level of structural complexity and patterns of embryonic development. They are often characterized by the type of symmetry demonstrated in the body plan and the number of tissue layers in the embryo.

Organ Systems

The simplest animals like sponges and cnidarians lack organ systems altogether, whereas structurally complex animals like mammals possess up to 11 different systems. Most animals have muscular, circulatory, digestive, and reproductive systems. Relatively fewer animals have the nervous, endocrine, immune, excretory, integumentary, respiratory, and skeletal systems.

Organs

Each organ system comprises individual organs working together for coordinated function. For example, the human digestive system contains openings (such as the mouth); tubes (such as the esophagus and intestines); and chambers (such as the stomach and rectum).

Tissues

Each organ is made up of different tissues, including nervous, muscle, connective, and epithelial. *Nervous tissue* is structured in such a way that it rapidly conducts electrochemical messages throughout the body to control behavior, thoughts, and movements. *Muscle tissue* is composed of overlapping filaments that can slide across each other to allow for shortening of the fibers (as in a contraction); thus, muscle tissue is instrumental in animal movement. *Connective tissue* ranges from the tough, dense bone to the

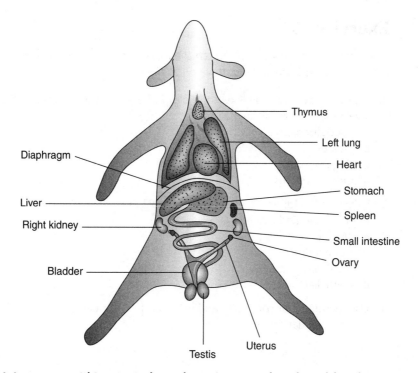

Diaphragm

Liver

Right kidney

Bladder

Thymus

Left lung

Heart

Stomach

Spleen

Small intestine

Ovary

Testis

Uterus

Major organs within a typical vertebrate (Note: Both male and female sex organs shown.)

loose, aqueous blood, with cartilage falling somewhere in between. *Epithelial tissue* provides protection, as it is structured to line the surfaces of organs. It can be *simple* (one layer), *stratified* (multiple layers), or *pseudostratified* (one irregular layer appearing as multiple layers). It is also characterized by the shape of the cells it contains, namely *cuboidal* (box-shaped), *columnar* (elongated), or *squamous* (flattened).

Cells

Each tissue is made up of specialized cells that are each the product of differentiation from *stem cells*. Embryos possess unique stems cells that are *totipotent*, or completely undifferentiated, and can potentially become any cell type in the body. The stem cells of adult organisms, on the other hand, are already somewhat differentiated, so they can only produce a group of related cell types. For example, the stem cells found in the bone marrow of vertebrates are capable of further differentiating into all types of blood cell, including red blood cells and a variety of white blood cells.

Exercise 16.1

If you can successfully work through the following questions, you should feel confident about being able to move on to the next section.

1. Which of the following levels of biological organization represents the highest level possessed by all animals?
   a. Specialized cells
   b. Tissues
   c. Organs
   d. Organ systems

2. Which of the following tissue types is most directly involved with the protection of other cells and tissues?
   a. Muscle
   b. Connective
   c. Nervous
   d. Epithelial

3. Use the theme of continuity and variation to describe the concept of stem cells and differentiation.

   _____

   _____

   _____

To successfully coordinate and control the various organs and organ systems within an individual animal, homeostasis is key. Each system is highly regulated through *feedback loops*, mechanisms for detecting when the levels of a particular substance are outside the optimal functional range and then responding accordingly to return conditions to normal.

Animal Physiology: Organ Systems

While the sponge represents those animals that lack true tissues, most animals possess tissues that are in turn organized into organs. Those organs are then coordinated structurally and functionally within organ systems. There are usually 11 different body systems described, and while all are found in humans and other mammals, many animals possess only some of these organ systems.

Movement: Muscular and Skeletal Systems

When present, the skeletal and muscular systems coordinate within an animal to produce movement. Most vertebrates possess a bony endoskeleton integrated with tissues like ligaments, tendons, and cartilage at the joints (regions where different bones meet). *Tendons* are specifically used to connect bones to the skeletal muscles, which are controlled voluntarily within the muscular system. These muscles are described as *striated* in appearance because of the thick myosin and thin actin filaments that permit contraction and relaxation. When signaled to contract, the muscles slide over each other in an overlapping fashion, resulting in shortening (contracting). The fibers release and lengthen during relaxation.

Other types of muscle, like smooth and cardiac, are not under voluntary control. Smooth muscle (named for its lack of striations) lines internal organs for the movement of substances through them. Cardiac muscle is striated like skeletal muscle, but it is involuntarily controlled and stimulated by its own pacemaker.

Homeostasis: Circulatory and Immune Systems

An animal's circulatory system is designed to move substances efficiently throughout the body in an aqueous medium. It is usually propelled by a muscular heart that forces blood through a vast network of arteries, capillaries, and veins. One of the major components of blood is *red blood cells*, those that bind to oxygen and deliver that life-sustaining substance to the body's cells for cell respiration and the production of ATP.

Other components of blood are the platelets and white blood cells, both of significant importance in immune function. The *platelets* are actually fragments of cells used in the blood-clotting process. In a coordinated sequence of steps, they block a laceration in a blood vessel. *White blood cells*, on the other hand, are highly specialized cells of the immune system. Some patrol the body in search for any foreign invader, but others are part of the animal's specific immune response. Some of these cells, like *B cells* and *T cells*, are designed to fight off one specific pathogen when it is detected inside the body.

Nutrition: Digestive and Respiratory Systems

An animal's digestive system usually involves the following steps: *ingestion*, *digestion*, *absorption*, and *elimination*. Depending on the animal's specific diet and ecological role, the exact components of a digestive system vary. In

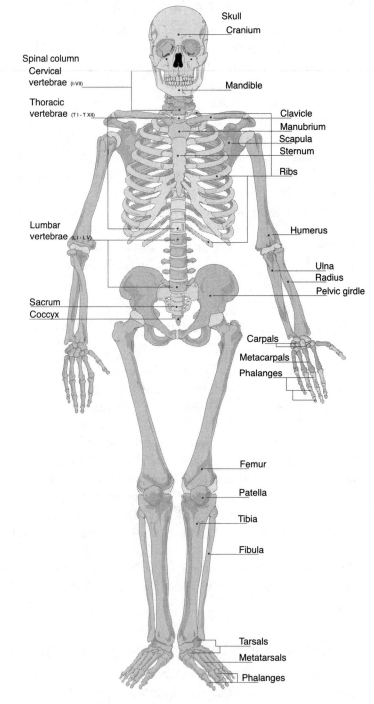

Skull
Cranium

Spinal column
Cervical
vertebrae (I-VII)

Mandible

Thoracic
vertebrae (T I - T XII)

Clavicle
Manubrium
Scapula
Sternum

Ribs

Lumbar
vertebrae (L I - L V)

Humerus

Ulna
Radius
Pelvic girdle

Sacrum
Coccyx

Carpals
Metacarpals
Phalanges

Femur

Patella

Tibia

Fibula

Tarsals
Metatarsals
Phalanges

The human skeletal system

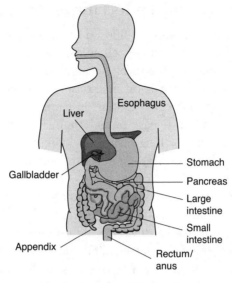

The human digestive system

general, however, the mouth is involved in ingestion; the mouth, stomach, and intestines play roles in digestion; the intestine specializes in absorption; and the anus allows for final elimination of wastes. Usually a combination of *mechanical digestion*, like the crushing and slashing action of the teeth and the churning of the stomach, and *chemical digestion* through specialized enzymatic action is required for optimal digestion within an animal.

The respiratory system is designed to pull oxygen gas into the body and move it to the bloodstream. (Remember that red blood cells are structured to bind to oxygen for cell delivery.) While not an energy-giving nutrient, oxygen is a necessary component for aerobic cellular respiration. Without it, animals would be considerably less energy-efficient and much less complex in structure.

The respiratory system differs quite markedly depending on habitat. An *aquatic animal* is likely to possess gills, featherlike structures that promote diffusion of dissolved oxygen from the water into the capillary beds surrounding them. *Terrestrial animals*, on the other hand, are likely to possess lungs. These delicate organs are positioned deep within the body to encourage the humid environment that gas exchange on land requires. Muscles are also used to pull air into the lungs from the atmosphere and then propel carbon dioxide wastes out after gas exchange has occurred.

Communication: Endocrine and Nervous Systems

The endocrine system is composed of a series of endocrine glands that produce and secrete hormones, chemical messengers in the body. These hormones are designed to travel through the blood to other target organs in the body. There, in a coordinated set of reactions, a desired outcome is triggered. The endocrine system is responsible for daily cycles like sleep (coordinated by the release of melatonin from the small pineal gland in the brain), as well as for long-term development like puberty (initiated by the release of either testosterone or estrogen from the male or female gonads, respectively).

The nervous system is designed much differently from the endocrine system for fast communication. It is composed primarily of neurons, elongated cells with tentacle-like *dendrites* on one end to receive information; a cell body that contains the nucleus and coordinates the response; and the long, thin *axon* down which the message travels. Each neuron operates by moving sodium and potassium ions across the membrane of the axon, thus transmit-

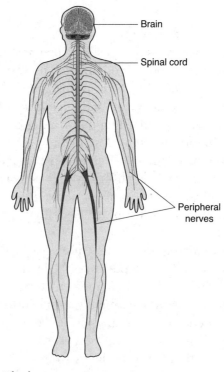

Brain

Spinal cord

Peripheral nerves

The human nervous system

ting an electrochemical message. When the message reaches the *synapse* at the terminus of the axon, it triggers the release of a neurotransmitter across that space. Receptors on the subsequent membrane—whether another neuron, a muscle, or a gland—pick up the message and initiate an appropriate response (like transmitting another signal, signaling contraction of a muscle, or encouraging a gland to secrete its hormone product).

Managing Wastes: Excretory and Integumentary Systems

The excretory system is primarily designed to eliminate metabolic wastes from the animal, usually from the bloodstream. In vertebrates, the kidney is the main organ of excretion. While the liver's metabolism of proteins is actually responsible for the production of the waste product urea, the blood transports the urea to the kidney for final removal. The kidney works under high pressure to force blood fluids into its internal *nephrons*.

These series of long, convoluted tubules carefully coordinate the exchange of materials between the nephron tubules and the surrounding capillary beds. The bottom line is that anything the body wants to eliminate as waste needs to end up in the nephrons, while anything the body wants to keep (mainly water and nutrients) must be moved back into the bloodstream. At the end of each nephron is a collecting duct that moves the now-concentrated urine to the bladder for storage and eventual removal from the body.

The integumentary system is composed of the outer coverings of the animal. The integument of humans, for example, is made up of skin, hair, and nails. These three substances are all keratinized to remain watertight, yet pores in the skin allow for the excretion of nitrogenous wastes through sweat and thermoregulation through evaporative cooling (sweating) and the shivering response. The skin is also considered a primary yet nonspecific immune organ, as it provides a covering that keeps the vast majority of pathogens and other potential threats out.

Reproduction: Male and Female Systems

Male reproductive systems are designed, once the animal reaches reproductive maturity, to produce and mature sperm cells. These cells, like all gametes, are haploid and thus are created through meiosis from the animal's diploid cells. Differentiation produces each sperm cell with a long flagellum for propulsion through an aqueous medium to reach and fertilize an egg.

Female reproductive systems are designed to both produce and mature female gametes (ova, or eggs) and to promote appropriate development of the embryo (assuming fertilization is internal).

The excretory system is often confused with the final stage of digestion, elimination. It is important to remember that if a substance is ingested but never absorbed, it was never actually part of the organism (it was really just passing through) and was eventually lost through elimination. Excretion involves the removal of waste products produced by the organism through metabolism or of any toxins to which the organism has been exposed in its environment.

Exercise 16.2

If you can successfully work through the following questions, you should feel confident about being able to move on to the next section.

1. Which of the two systems paired in the list below do *not* interact directly?
   a. Nervous and muscular
   b. Respiratory and circulatory
   c. Skeletal and muscular
   d. Respiratory and immune

2. Through what means are the reproductive and endocrine systems related?
   a. Nervous signaling
   b. Hormone action
   c. Enzyme catalysis
   d. Excretory function

3. Use the theme of regulation: feedback and response to describe the concept of excretion and kidney action.

   _____

   _____

   _____

17

Populations and Communities

☑ *Ecology* is the study of organisms and the nonliving world they inhabit.

☑ Central to ecology is the complex set of interactions between organisms, both *intraspecific* (between members of the same species) and *interspecific* (between members of different species).

☑ *Population ecologists* and demographers study population density, dispersion, and dynamics. They use population growth models to understand current and predict future population changes.

☑ Interspecific interactions are studied by *community ecologists*. These include predator-prey interactions, interspecific competition, and a variety of symbiotic relationships.

☑ *Ecological succession* describes the typical patterns of reestablishing populations observed within a community after a major environmental disturbance.

Ecological Levels of Organization

At the fundamental level, ecology should be approached from the perspectives of increasing levels of biological organization. Because at its most focused level of study, ecology investigates the interactions between individuals, population is the lowest organization level of significance.

A Population

A group of organisms of the same species living together in the same geographic area constitutes a *population*. For example, all of the mockingbirds living on the island of Hood in the Galapagos are members of the same population. The mockingbirds on the neighboring island of San Cristobal, while closely related, are actually members of a different species, because the birds cannot and do not interbreed. *Intraspecific interactions*, those between members of the same species, are of concern to population ecologists.

A Community

A community is the next level of biological organization just after population. A *community* includes all of the various species living together within a given habitat. Revisiting Hood from the previous scenario, the mockingbirds along with all of the other organisms—whether fungi, plants, or other animals—that occupy the same regions of the island as the mockingbirds and that interact directly or indirectly would constitute a biological community. *Interspecific interactions*, those between members of different species, are of concern to community ecologists.

An Ecosystem

All of the biological communities within a given geographic area make up an *ecosystem*. At this level of organization, nonliving factors (called *abiotic*) become relevant. These include elements like ambient temperature, precipitation levels, and soil quality. Many biotic factors are also significant, including the types of pathogens present, predator population levels, and the effects of human activity.

A lake ecosystem

A Biome

A *biome* is a large, sweeping geographic area made up of many interconnected ecosystems. An individual biome is characterized by a typical climate and by biological communities with similar adaptations. Most terrestrial biomes (such as tropical rain forests and deserts) are well studied, but both freshwater and marine aquatic ecosystems are less understood due to the practical challenges of studying underwater life.

The Biosphere

The composite of all the biomes that exist on Earth to support life is called the *biosphere*. At its essence, the biosphere is the living portion of the planet. For now, life as we recognize it exists only on Earth, but ongoing research seeks to discover any forms of life on other planets and in other galaxies.

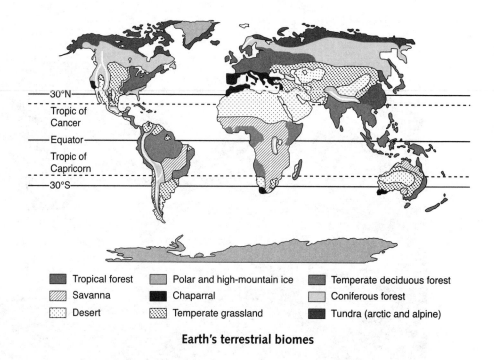

Earth's terrestrial biomes

 The notion of life on other planets has often been exaggerated and sensationalized. Much research is ongoing to discover whether Mars has—or more likely, ever did have—any form of life inhabiting its surface. If so, this organism would likely be similar to some of the earliest life-forms that inhabited our own planet—*extremophiles* like anaerobes.

Exercise 17.1

If you can successfully work through the following questions, you should feel confident about being able to move on to the next section.

1. Which of the following levels of ecological organization is the least inclusive?
   a. Community
   b. Biome
   c. Biosphere
   d. Population

2. Which of the following are significant emergent factors at the ecosystem level of biological organization?
   a. Abiotic factors
   b. Biotic factors
   c. Population levels
   d. Predator-prey interactions

3. Use the theme of interdependence in nature to describe the concept of interspecific interactions.

_____

_____

_____

Populations and Intraspecific Interactions

The specific space in a natural environment that an individual organism occupies for survival and reproduction is called its *habitat*. The lifestyle of a particular species within its habitat is referred to as its *niche*. Members of different species must occupy different niches within a given community in order to allocate resources efficiently, resulting in interesting interspecific interactions (discussed later in this chapter).

Population Density

Population density is a measure of the number of individuals in a population relative to the geographic area those individuals inhabit. In other words, population density is a measure of how crowded individuals are within their habitat. If population density creeps too high, then limiting factors in the environment become increasingly important in affecting the welfare of individual organisms. For example, limiting factors for a population of bacteria living in a Petri dish in a lab include the size of the dish and the availability of essential nutrients supplied in the medium.

Population Dispersion

Population dispersion describes the pattern of distribution of individual organisms throughout their geographic range. Very few populations actually exhibit random distribution; most demonstrate uniform distribution or clumped distribution. Often, *uniform distribution* is associated with behaviors that influence such a pattern. For example, the nesting sites of certain

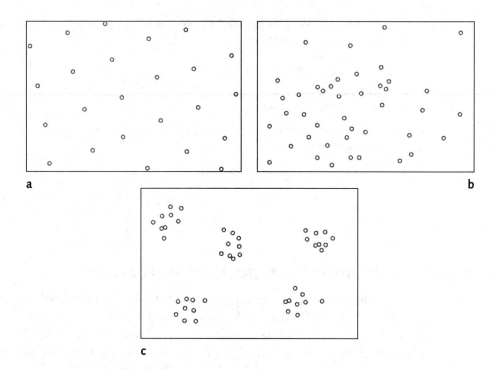

Three population dispersions: (a) uniform distribution, (b) random distribution, and (c) clumped distribution

birds are often distributed uniformly as a means of preserving social order in a crowded space. *Clumped distribution*, on the other hand, is often associated with individuals choosing or needing to live close to particular essential resources, or it is observed in individuals that live in natural groups or herds.

 ## Population Dynamics

Factors such as life expectancy, birthrate, and death rate are significant to *demographers*, those who study population growth and describe how the basic makeup of the population changes over time. Numbers of individuals immigrating to or emigrating from the population are also relevant. Demographers assemble graphic displays like *survivorship curves* and *age structure diagrams* to demonstrate the likelihood that individuals will survive to a certain age and the distribution of individuals represented at different age ranges, respectively.

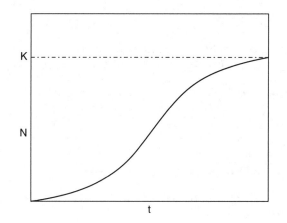

The change in population (N) over time (t) as it reaches carrying capacity (K)

Population Growth

Two basic models describe the growth of populations: exponential and logistic. *Exponential growth* is relatively rare, as it results from a continuous, steady increase in the population size over time. On a graph, this appears as a J-shaped curve. *Logistic growth* is observed more often; this type of growth is eventually slowed and limited by the carrying capacity (K) of the population. This produces an S-shaped curve on a graph.

The *carrying capacity* is determined by the limiting factors in the environment that most severely affect reproduction and/or survival. Some of these factors are described as density-dependent, indicating that their significance depends on the current population numbers. For example, the health effects of an infectious pathogen on a population depend on the number of host organisms that are interacting and the quality of such interactions. Other factors are described as density-independent, including abiotic factors like flash floods and heat waves.

Exercise 17.2

If you can successfully work through the following questions, you should feel confident about being able to move on to the next section.

1. Extremely territorial animals will most likely exhibit what form of dispersion pattern?
   a. Random
   b. Clumped
   c. Uniform
   d. Unknown from the information provided

2. Which of the following is most correctly associated with an S-shaped growth curve?
   a. Growth limited by the carrying capacity
   b. Bacterial growth
   c. Stable, steady growth
   d. Unregulated growth

3. Use the theme of interdependence in nature to describe the concept of density-dependent limiting factors.

   _____

   _____

   _____

Communities and Interspecific Interactions

Interspecific interactions, those between members of different species (and thus different populations) within a community, are central to community ecology. These interactions are examples of emergent factors at the community level of biological organization, for they are not relevant at the population level when only one type of species is being studied.

Predator-Prey Interactions

One of the most significant types of interspecific interactions in a community is that of predator-prey (animals actively consuming other organisms). Adaptations that evolve among the prey population, for example, include physiological and behavioral mechanisms that make it more challenging for the predator to find and/or consume them. For instance, common prey adaptations include *camouflage* (blending into the environment), *mimicry* (appearing to be something else that is undesirable), and *biotoxicity* (produc-

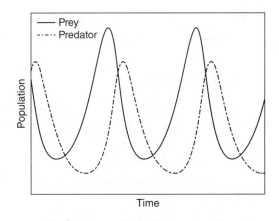

Predator-prey interactions over time

ing toxic substances harmful to potential predators). Often, biotoxic animals also show warning coloration, using bright body color to discourage future predation.

Given that the predator population is ultimately dependent on the prey population, the patterns of predator population growth and decline typically mirror yet lag those of the prey population. For example, as the gazelle population declines, the population of lions eventually declines too. This puts less pressure on the gazelle population, which slowly increases. This eventually supports a larger population of lions, which eventually influences a decrease in the gazelle population once again.

Interspecific Competition

Members of different species often compete for limited resources in an environment, as observed when two species of plants fight for sunlight in the thick canopy of the tropical rain forest. Sometimes shared resource use results in reduced populations of both species. Often, however, one species is much more efficient at obtaining and/or using the resources, resulting in the eventual elimination of the less-efficient species from the community. The competitive exclusion principle is used by community ecologists to describe this effect when observed in nature.

Other effects of interspecific competition include character displacement and resource partitioning. *Character displacement* results when natural selection has acted on populations within a competitive community in such a way as to produce traits that use resources differently. *Resource partition-*

ing describes how a species may avoid competition by using only a specific portion of a resource, leaving the remainder for other species to share.

Symbiotic Relationships

Understanding symbiotic relationships is very important to the study of community ecology. Symbiotic relationships are generally described as taking one of three forms. The most common is *mutualism*, a relationship between organisms that is beneficial to both participants (+/+); examples include lichens (fungi and unicellular photoautotrophs) and flowering plants and their pollinators. Other relationships, like that between a bedbug and its human host, exhibit *parasitism*, because they benefit one participant while harming the other (+/−). Finally, some relationships demonstrate *commensalism*. They are beneficial to one participant while neither helping nor harming the other (+/0). Examples of commensalistic relationships are less common in nature and include barnacles residing in the flippers of whales and birds inhabiting the natural notches within a tree.

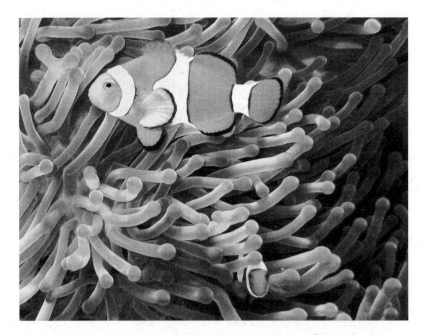

A mutually beneficial symbiotic relationship between clownfish and sea anemones

Species Richness and Species Evenness

Measures of the health and stability of a community include species richness and species evenness. *Species richness* is a measure of the number of different species present within a given community. From the global perspective, species richness increases with movement from either pole toward the equator. On a smaller scale, it is a useful measure in understanding the sensitivity of a community to a potential ecological disturbance.

Species evenness is a measure of the relative abundance of each species within the community. It is, in essence, a measure of a community's biodiversity. It can provide a sense of the balance of ecological roles within an environment (such as producers and consumers).

Community Disturbances and Succession

When a biological community faces a major environmental disturbance, typical patterns of reestablishing populations are observed. If the disturbance is so severe that it results in completely new sediment on which new plant communities must then be established, then the succession is described as primary. *Primary succession* results after events such as a volcanic eruption that covers original sediment in lava or a glacial movement that removes all preexisting layers of productive soil. It proceeds very slowly and is characterized by the initial establishment of *pioneer species*, those that typically are the first to take up residence and establish a new community.

However, if the soil is left intact after a disturbance, then the succession is secondary. *Secondary succession* is observed in nature much more frequently than primary succession; it is generally seen after a forest fire or flash flood.

Many symbiotic relationships are often misclassified as commensalistic when they are, in fact, mutualistic. Upon close investigation, an organism usually gains some benefit, even if it is relatively small, by interacting with another.

Exercise 17.3

If you can successfully work through the following questions, you should feel confident about being able to move on to the next section.

1. In which of the following is the interspecific interaction mismatched with a result?
   a. Predator-prey—prey camouflage
   b. Interspecific competition—resource partitioning
   c. Interspecific competition—character displacement
   d. Commensalism—biotoxicity

2. Which of the following is associated with secondary succession?
   a. New soil in which to establish plants
   b. Very slow growth of populations
   c. Pioneer species
   d. Repopulation from dormant seeds

3. Use the theme of interdependence in nature to describe the concept of predator-prey population dynamics.

   _____

   _____

   _____

18

Ecosystems and Biomes

☑ Ecologists categorize significant factors in an ecosystem as either abiotic (non-living) or biotic (living).

☑ Biogeochemical cycles show how nutrients are recycled through an ecosystem, while an energy pyramid demonstrates the flow of energy through organisms in an ecosystem.

☑ Food chains and food webs help illustrate the specific means by which nutrients and energy are transferred between organisms in an ecosystem.

☑ On a larger scale, biomes represent the collection of many different, inter-related ecosystems that have characteristic inhabitants and typical climatic patterns. Terrestrial biomes are much better described and understood than aquatic ecosystems.

Ecosystems

Recall that an ecosystem is an assemblage of all of the organisms making up the biological communities in a designated environment, as well as the abiotic factors present. Certain geochemical cycles link all ecosystems and explain the means by which essential nutrients are effectively recycled as they are passed between organisms.

Biogeochemical Cycles

Four major biogeochemical cycles are of significance to ecologists, namely the carbon, water, nitrogen, and phosphorus cycles. The *carbon cycle* is central to biology given that organisms are built from and dependent on carbon. The primary way in which carbon is transferred between organisms is through the dual gas exchange processes involved within photosynthesis and cellular respiration. As autotrophs absorb carbon dioxide (CO_2) to build glucose, they remove the same waste product that heterotrophs release into the atmosphere through cell respiration. When heterotrophs consume the autotrophs, they gain access to carbon for their own metabolism. Today's carbon cycle on Earth is vastly out of balance, as combustion of fossil fuels and other

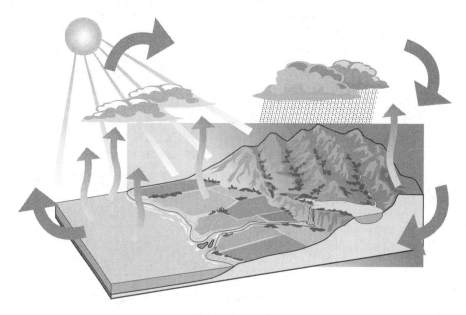

The water cycle

products of human industry introduce more CO_2 into the atmosphere, while through deforestation, humans remove organisms that absorb CO_2.

The *water cycle* is also important in understanding how organisms have ready access to that ever-important inorganic resource. Organisms give off water vapor through many processes (consider transpiration in plants or respiratory condensation in animals) and also lose liquid water through evaporative cooling (sweating) or excretion of nitrogenous wastes in urine. Animals obtain water through drinking and eating, as well as by creating it metabolically during anabolism. Plants and fungi obtain it through absorption from the roots and hyphae, respectively.

The *nitrogen cycle* helps scientists understand how organisms recycle the necessary nitrogen for proteins and nucleic acids. Soil bacteria are primarily responsible for most conversions within the cycle, as they initially convert atmospheric nitrogen gas into an ionic form such as a nitrate (NO_3^-) or a nitrite (NO_2^-) that a plant can then absorb as it takes up water through its roots. Other nitrogenous waste products from excretion and decomposition are returned to the soil as well, and other bacteria participate in converting them to the ammonium ion (NH_4^+).

The *phosphorus cycle* similarly helps explain how the phosphorus that is so crucial for nucleic acid structure is recycled through an ecosystem. It involves more geologic processes (such as the weathering of rocks) than biological ones (predominantly decomposition) that produce the phosphate ion (PO_4^{3-}). This ionic form is then readily absorbed by plants and fungi and ingested by the animals that consume them.

Trophic Levels and the Energy Pyramid

Organisms within an ecosystem are organized energetically into trophic levels by their positions within the larger energy transfer process. The *primary producers* are placed at the bottom of the pyramid; they are usually photosynthetic (although chemosynthetic bacteria may constitute this level in specialized ecosystems) and possess the most biomass of any of the trophic levels overall. The next trophic level is made up of *primary consumers* (mainly herbivores) that have less biomass than the producers on which they depend. As the primary consumers eat the producers, they obtain energy and nutrients. The pattern continues with subsequent trophic levels (secondary consumers, tertiary consumers, and so on).

Nutrients are passed efficiently from organism to organism, but a great deal of energy is lost in the process, meaning that limited sequential transfers can occur. When trophic levels are placed on top of each other in order

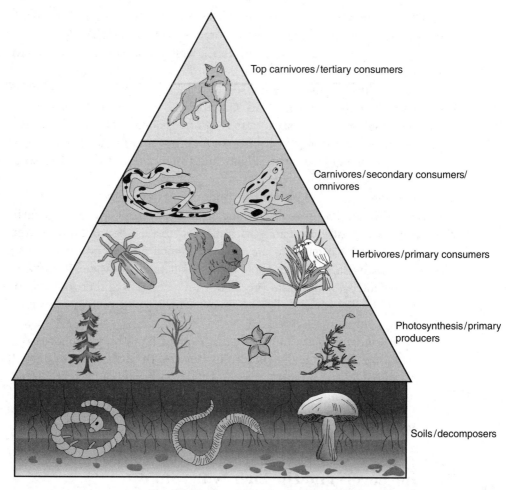

Trophic levels within an energy pyramid

of energetic dependence, they taper with each level, creating an energy pyramid. Due to the 10 percent rule, which states that the maximum level of net energy transferred to the next trophic level is 10 percent, the highest level of consumer possible (but not always present) in any natural ecosystem is typically quaternary (fourth level).

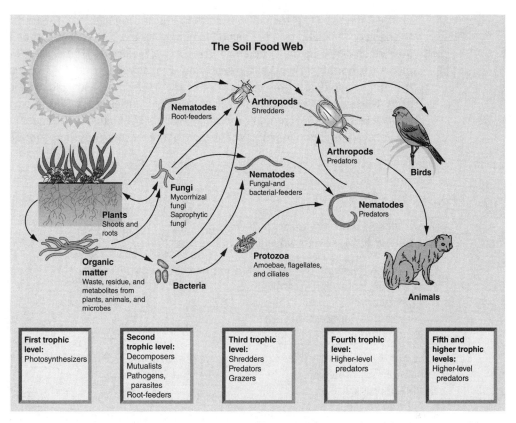

Typical food web within a soil ecosystem

Food Chains and Food Webs

A *food chain* is a specific sequence of energy transfer steps that are possible between successive trophic levels within a given ecosystem. It begins with a primary producer (such as a lava cactus) and ends with a top-level carnivore (such as a Galapagos hawk). When the intersection of the various food chains within an ecosystem is examined in its composite form, it is called a *food web*.

When studying the concept of the energy pyramid, it is easy to assume that the trophic level just above the primary producers comprises the secondary consumers because of the sequential-sounding nature of their names. Remember that there is only one level of producers, and it is always situated on the bottom of the pyramid. The next level up must then be the first-order, or primary, consumers. Secondary consumers and so on follow.

In attempting to keep track of the mechanisms and significance of each of the biogeochemical cycles, it is helpful to remember the CHON elements and their significance in the different organic macromolecules. The CHO elements (carbon, hydrogen, and oxygen) are fundamental to all four of the macromolecule types, from carbohydrates and lipids to proteins and nucleic acids. Nitrogen (N) is common only in proteins and nucleic acids. Phosphorus is much less common in organic molecules overall, although it is abundant in nucleic acids as a fundamental part of each monomer (the phosphate group).

Exercise 18.1

If you can successfully work through the following questions, you should feel confident about being able to move on to the next section.

1. Which of the following trophic levels is mismatched to its description?
   a. Secondary consumer—carnivore
   b. Primary producer—autotroph
   c. Tertiary consumer—carnivore
   d. Primary producer—omnivore

2. Which of the following cycles is primarily responsible for recycling nutrient components of proteins?
   a. Phosphorus cycle
   b. Nitrogen cycle
   c. Carbon cycle
   d. Both a and c

3. Use the theme of continuity and variation to describe the current imbalance of the carbon cycle on Earth.

   _____

   _____

   _____

Terrestrial and Aquatic Biomes

Recall that biomes are large geographic ranges with characteristic climatic patterns and biodiversity. Terrestrial biomes, like the deciduous forests and savannas, are well described; aquatic biomes, especially the marine biomes and the deep sea in particular, are less so. Until recently, when deep-sea submersible technology became available for research, biologists knew very little about what existed far out of the reach of scuba divers and traditional submarines.

Tundra

The *tundra* is the northernmost biome, characterized by very low annual temperatures and relatively little precipitation. In terms of biotic composition, it typically has sparse, small vegetation and no large shrubs or trees. This is in large part due to its permanently frozen soil called *permafrost* that lies just beneath the upper soil.

Taiga (or Coniferous Forest)

The biome just to the south of the tundra is typically *taiga*, or coniferous forest. It is characterized by low temperatures and significant precipitation. The dominant plant species are usually conifers, trees bearing cones and needle-like leaves. These are well adapted for the nutrient-poor soil and the large amounts of snow that would potentially collect on broad leaves.

Deciduous Forest (or Temperate Broadleaf Forest)

The *deciduous forest* often borders the taiga just to the south. Its soil is richer in nutrients, and it experiences more distinct seasons. A moderate amount of rainfall is typical in the deciduous forest, as are broadleaf trees that lose their leaves in the winter when the sun's intensity is not enough to support them. Conifers and evergreens are also common there.

Grasslands

Different grassland biomes exist, including temperate grasslands, savannas, and chaparrals. The *temperate grasslands* typically grow at the same latitude as the deciduous forests, but they are located toward the interior of continents instead of near the coastlines. The relatively low levels of precipitation do not support tall trees, so thick grasses and shrubs dominate instead. In the *savannas*, some tall trees and large shrubs are scattered throughout the grassland, supported by the annual rainy season. The grasses of both temperate grasslands and savannas support large herds of grazing animals. *Chaparral* is a grassland-desert hybrid biome that is limited to the coastal regions of the Mediterranean and Southern California and characterized by a very mild climate.

Deserts

Deserts are well known for their very low levels of annual precipitation. Although often assumed to be hot, many deserts experience extremes in temperature, with relatively low temperatures at night. Desert vegetation is typically sparse and rich with species, such as fleshy cacti and succulents, that have specially adapted for such an extreme climate.

Rain Forests

Tropical *rain forests* are often described as the most biodiverse terrestrial biome. Positioned near the equator, they are characterized by warm temperatures, consistent sunlight intensity, and high levels of precipitation throughout the year. They are known for the thick canopy of tall trees that results, many of which are coated in smaller plants called *epiphytes*, which have adapted to grow in the upper branches without ever rooting into the soil.

Aquatic Biomes

While ecologists are more familiar with terrestrial biomes, aquatic biomes cover a much larger proportion of the Earth's surface (approximately 70 percent). The extreme depths of the ocean (average depth is more than 3.5 kilometers) also contribute to a massive volume of space within which marine organisms can live. Marine ecologists divide the oceans into zones; both the distance from the shore and the distance beneath the ocean's surface are significant in describing the zone. Generally speaking, organisms that occupy the upper levels of the water table and have access to the sun's radiant energy inhabit the *photic zone*. Those living beneath that level occupy the *aphotic zone*. Organisms living on the ocean floor, whether photic or aphotic, are called *benthic*.

The photic zone that is nearest the shoreline and is partially exposed to air on a daily basis with the ebb and flow of the tides is called the *intertidal zone*. Just past the intertidal zone is the *neritic zone*, which stretches out as far as does the continental shelf. It is the most productive part of the ocean, rich in nutrients and home to coral reefs. Past the continental shelf lies the *pelagic zone*, or open ocean. The aphotic depths of the open ocean are called the *oceanic zone*. Some of the least understood biomes exist here, including deep-sea vent communities that are ultimately supported by chemosynthetic bacteria.

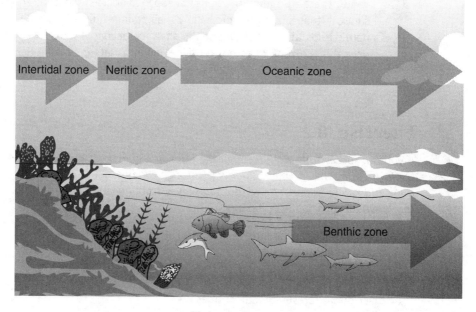

Main ocean zones

Freshwater biomes include moving bodies of water like rivers and streams, sedentary bodies like lakes and ponds, and transitory bodies like freshwater estuaries. Rivers and streams are bodies of freshwater that move down a slope toward a mouth. Murky, slow-moving rivers tend to be richer in nutrients and are thus more supportive of life than rocky, fast-moving rivers with clear water. Most lakes are described as *eutrophic*, also murky in appearance due to the lush vegetation and nutrient-rich waters. Fewer lakes are characterized as *oligotrophic*, possessing clear waters with rocky bottoms and supporting relatively little organic matter.

Benthic does not necessarily imply aphotic, although this is an easy mistake. Remember that *benthic* just means "bottom-dwelling." Tide pool communities in the intertidal zone and coral reef communities are examples of areas that are both benthic and photic.

It can sometimes seem overwhelming to keep track of the various ocean zones. There are a few conditions to keep in mind. First, does the zone receive sunlight (that is, is it photic)? If so, it may be intertidal if it's at the shoreline, neritic if it's above the continental shelf, or pelagic if it's part of the open ocean. If the zone does not receive sunlight (that is, it's aphotic), then it's likely part of the oceanic zone.

Exercise 18.2

If you can successfully work through the following questions, you should feel confident about having made your way through the final set of questions in the book and gaining a much more thorough understanding and appreciation of the fascinating discipline of biology.

1. Which of the following terrestrial biomes is matched with the *incorrect* characteristic?
   a. Taiga—cone-bearing trees
   b. Temperate grassland—epiphytes
   c. Deciduous forest—broadleaf trees
   d. Tundra—permafrost

2. Which terrestrial biome is characterized as having the highest degree of species richness?
   a. Tundra
   b. Temperate grassland
   c. Tropical rain forest
   d. Desert

3. Use the theme of the relationship between structure and function to describe the concept of eutrophic lakes and slow-moving rivers.

   _____

   _____

   _____

Answer Key

Chapter 1
The Study of Life

1.1   1. c   2. a   3. A cell is a complex unit containing specific structures that enable it to carry out different functions. For example, the cell membrane surrounds the outer surface of a cell and therefore can provide a protective function in preventing or permitting the crossing of the surface by substances.

1.2   1. a   2. d   3. Organs must work together within a given organ system in order to function appropriately. There must be continuity between the organs so that they communicate and coordinate effectively, but each has its own unique structure that allows it to carry out its specialized function. If all organs function appropriately, then the organ system functions appropriately too.

1.3   1. b   2. c   3. Metabolism is the sum of all chemical reactions that occur within a cell or an organism. Whether molecules are being built up or broken down, energy transfer takes place with each chemical reaction through the breaking and formation of chemical bonds.

1.4   1. b   2. d   3. The microscope is an essential tool used by biologists to study life that is too small to be seen with the naked eye. This technology allows biologists to gain knowledge about the microscopic world so they can apply it to other fields, such as biomedicine (treating viral and bacterial diseases, and so on).

Chapter 2
The Chemistry of Life

2.1   1. c   2. c   3. Using water as an example, it becomes clear how energy transfer relates to the three states of matter. When energy is added to a block of ice, the ice will eventually turn into liquid water. If more energy is added, the liquid water will evaporate and create water vapor.

2.2   1. c   2. a   3. As organic monomers are joined together through chemical bonds, polymers are created. The digestion of organic polymers releases the monomer building blocks to be used again (either by that same organism or by another organism that consumes the first).

2.3   1. a   2. d   3. Carbohydrates represent short-term and intermediate-term energy for a cell. Monosaccharides are ready energy for the cell, because they can be converted directly into ATP. Excess monosaccharides are stored in complex polysaccharide form, such as starch or glycogen. Long-term energy is provided by lipids, whose nonpolar fatty acid chains are much tougher to break down.

Chapter 3
Cell Structures and Functions

3.1   1. a   2. a   3. Under certain conditions, unicellular organisms have been known to group together to form a colony. The collective group of organisms thus depends on the overall functioning of the colony for survival. Over time, certain mechanisms can increase that interdependence, eventually making the individual cells unable to function on their own.

3.2   1. b   2. d   3. DNA stores a cell's genetic information, which ribosomes eventually translate into proteins. If the sequence of a gene is changed (mutated), then the ribosomes work from erroneous instructions and build an incorrect protein. If the protein is not structured correctly, then it likely will not function effectively.

3.3   1. c   2. d   3. The endomembrane system is a series of membranous surfaces within a cell that is continuous either by its structure (for example, the ER often branches off of the nuclear envelope) or through the action of vesicles. As the rough ER produces a protein, it uses its own membrane to create a vesicle to surround the protein and direct it to the Golgi. There, the vesicle membrane fuses with the Golgi's membrane for further processing and packaging.

Chapter 4
Cell Homeostasis and Metabolism

4.1   1. b   2. c   3. The cell membrane is largely responsible for regulation of what enters and leaves a cell. It is well suited for this purpose due to its selective permeability. The thick, hydrophobic interior of the membrane is due to tightly packed hydrophobic tails that act as a physical and chemical screen to prevent ions and most polar substances from passing.

4.2   1. b   2. d   3. The light reactions involve the trapping of light energy by pigments deep within the chloroplasts that kick-start a series of reactions. Eventually, some energy is stored in the form of ATP and NADPH. During the dark reactions, this ATP and NADPH energy is invested to use CO_2 to build glucose.

4.3   1. d   2. c   3. Glucose is very stable, so energy must first be added to make it less stable and to initiate the breakdown. Once two ATP molecules have been used to phosphorylate glucose, the molecule breaks down slowly and eventually creates pyruvate. Pyruvate can then to be broken down by either fermentation or the Krebs cycle and the ETC to extract as much remaining ATP as possible.

Chapter 5
The Cell Cycle and Cell Division

5.1   1. c   2. b   3. The G_1 checkpoint is a clear example of regulation in that it determines whether a cell will continue on in the cell cycle and eventually divide, or whether it will stay in a "holding phase" and not divide unless it receives further instruction. This fundamental process prevents cancer from occurring much more frequently.

5.2   1. b   2. c   3. In chromatin form, DNA is loosely arranged and disorganized. In this form, it can be read for processes such as replication and transcription. When DNA is condensed and wrapped into chromosomes, the genetic code within the DNA molecule is largely inaccessible. However, the chromosomes can be more easily managed for the process of mitosis.

Chapter 6
Heredity: Meiosis and Mendelian Genetics

6.1   1. b   2. d   3. Haploid cells contain half the genetic material that a typical cell for that species contains. In the context of animal reproduction, the haploid cells are intended to fuse with another gamete during the fertilization process, and the two halves combine to make a whole (a diploid cell). A diploid cell has two sets of chromosomes, one inherited from each parent.

6.2   1. a   2. d   3. Gregor Mendel repeated his cross-pollination experiments with pea plants many times and used statistics to analyze his data. The information gained from his work (Mendel's laws of inheritance) has shaped all current knowledge of genetics and heredity.

Chapter 7
DNA and the Genetic Code

7.1   1. d   2. a   3. When DNA is replicated, there are efforts to achieve perfect continuity of information (such as DNA proofreading and repair mechanisms). If perfect continuity is achieved, offspring cells will have exactly the same instructions as the parent cell, and any proteins produced through transcription and translation will have the correct sequence of amino acids. If a mutation (variation) occurs, the gene code changes, and any proteins produced will contain incorrect information and likely be dysfunctional.

7.2   1. d   2. c   3. To ensure that replication takes place accurately, DNA polymerase first proofreads the complementary base pairing. If an incorrect nucleotide has been placed erroneously, it is replaced by the appropriate one. After replication, another enzyme rereads the new molecule and corrects any mistakes. Still other enzymes scan DNA and correct any mistakes that accumulate over time.

7.3   1. c   2. d   3. mRNA is made by transcribing a gene. It is maintained as a linear molecule so it can be read easily during translation by a ribosome. tRNA is shaped like a clover leaf, with an anticodon at one end and the corresponding amino acid at the other. The anticodon can easily interact with the mRNA codon in this configuration, and the amino acid is easily accessible for dehydration synthesis with the growing polypeptide chain. rRNA is globular, which provides the ribosome with a three-dimensional structure to support the interaction between mRNA and tRNA.

Chapter 8
Human Genetics and Biotechnologies

8.1   1. c   2. d   3. A karyotype is a display of chromosomes created from arresting a cell during metaphase in mitosis and then arranging the chromosomes from largest to smallest, with the sex chromosomes positioned last. The karyotype can then be used to assess the biological sex of the individual, whether there are any extra or missing chromosomes, or whether the individual has significant chromosomal mutations.

8.2   1. c   2. c   3. If the structure of a chromosome changes in a significant way (such as through a mutation like translocation), then its function is also affected. In the case of translocation, one chromosome loses a stretch of genes and thus the ability to make the proteins coded for in those lost genes.

8.3   1. a   2. b   3. Reproductive cloning involves taking the genome from an organism that already exists and using that information to make a new organism. The idea is for the cloned organism to be an exact copy of the parent, so no genetic recombination takes place as it would when natural fertilization occurs for reproduction.

Chapter 9
The Origins of Life on Earth

9.1   1. d   2. c   3. Biogenesis involves the production of new life from existing life, as observed during asexual and sexual reproduction. If asexual reproduction occurs, then one parent passes on an exact copy of its genome to the new offspring, and there is no variation. If sexual reproduction occurs, then two parents each contribute half of their genome to their offspring, so variation occurs between generations.

9.2   1. b   2. d   3. As precells became increasingly complex and spontaneously assembled useful information that helped them survive, the most successful precells outcompeted the less successful. Over time, only those cells that contained genetic information could pass on that successful structure to subsequent generations.

9.3   1. b   2. d   3. If one cell engulfs another and the now internal cell maintains its original structure, then an interdependence can arise. In the case of endosymbiosis, the internal cell over time became an organelle that specialized in structure and function, so the larger cell that was originally prokaryotic had internal membrane-bound organelles and thus became eukaryotic.

Chapter 10
Evolution and Adaptation of Life

10.1   1. c   2. c   3. Evolution is the change in the genetic composition of a population over time. Usually genetic information remains constant from generation to generation, but genetic recombination and mutation can lead to novel genes. If these novel genes are beneficial to the organism, they contribute to the evolution of that species.

10.2   1. c   2. b   3. Homologous structures are those present in different species that were inherited from a common ancestor. The genetic information may have changed slightly through evolutionary processes, so the structures themselves may appear different in each species. However, the underlying structure and its genetic foundation are from the same source.

10.3   1. a   2. d   3. An evolutionary biologist may find it helpful to calculate gene and allele frequencies in a gene pool. By making this statistical determination, the scientist can assess whether evolution is actually taking place in the population over time. If the allele frequencies are changing, then evolution is occurring.

Chapter 11
Classification of Life

11.1   1. c   2. d   3. Taxonomy is the systematic process of naming and classifying organisms. All information known about the species is used to group it appropriately along with its close relatives. Being able to organize and manage the Earth's millions of species is essential in trying to understand life from the human perspective.

11.2   1. d   2. d   3. Given that animal cells lack cell walls, they also lack some of the protective structure that other cells possess. The advantage gained by this missing feature far outweighs its shortcomings, however. The ability of animal cells to communicate and coordinate more quickly enables nervous function and quick movement that is impossible in other organisms.

Chapter 12
Microbes: Viruses and Bacteria

12.1   1. b   2. c   3. A bacteriophage transfers its genetic information to a bacterial cell by using its structure to move between hosts. When it docks on the new host cell's surface, it inserts its genetic information through the cell wall and membrane into the host's cytoplasm. There, copies of the viral genome and new viral proteins can be made through replication and protein synthesis, and new viruses can be assembled. The cycle of infection continues from there, as the host cell bursts and releases its viral contents.

12.2   1. c   2. c   3. Cyanobacteria represent the major prokaryotic group of photosynthetic organisms. Especially early in evolutionary history, they would have played a very significant role in the ecosystem. Since they produced their own food, they not only gained an advantage themselves but also provided a food source for other organisms to consume as part of the food chain.

Chapter 13
Protists and Fungi

13.1   1. d   2. b   3. As they are currently classified, protists represent the most diverse kingdom of life, because they are classified primarily by exclusion from other kingdoms and not because of a shared set of common features. For that reason, many taxonomists seek to break up the protist kingdom into two or more appropriate kingdoms. This reworking of taxonomic groups can help biologists better represent and understand the evolutionary history of organisms and the relatedness between them.

13.2   1. c   2. b   3. Fungi can be symbiotic in two main ways. In a mycorrhiza, plant roots interact with the hyphae of soil fungi in a mutualistic fashion. The plant root effectively gains surface area, while the fungus gains food energy. A lichen also represents a mutualistic relationship, but in this case, the larger fungus provides protection and a habitat for the internal algae or cyanobacteria. The autotrophic inhabitant provides energy for the fungus in the form of glucose.

Chapter 14
Plants and Animals

14.1   1. c   2. d   3. When angiosperms are pollinated by animals like insects and birds, the animal gains nutrition through nectar and/or pollen from the plant. The animal then unknowingly transfers some of the pollen to the next plant it visits, thus contributing to the reproductive success of angiosperms via cross-pollination.

14.2   1. a   2. d   3. For the most part, the animal kingdom has demonstrated a shift from asymmetry to radial symmetry to bilateral symmetry throughout its evolutionary history. When the body style did not suit a particular species, a return to radial symmetry occurred (for example, echinoderms).

14.3   1. b   2. b   3. Endothermy allows an animal to regulate its internal body temperature despite changes in its external environment. Special receptors in the animal's surface covering initiate a response, like the shivering mechanism observed in humans. This returns a cooling body temperature back to normal and encourages optimal functioning of the organism.

Chapter 15
Plant Structure and Function

15.1   1. b   2. d   3. Plant organs consist of the roots, stems, leaves, and any specialized organ that is derived from one of these basic types (such as a flower). The structure of the organ dramatically influences its function, as is illustrated by plant leaves. A typical leaf is flat, broad, and thin, which makes it more effective for trapping sunlight for photosynthesis.

15.2   1. a   2. b   3. Guard cells regulate whether the stomata they surround are open or closed at any moment. They react to water levels internally and respond accordingly. If water levels drop, the guard cells wither and close the stoma to prevent further water loss. If water levels are high, the cells expand and open up the stoma, potentially allowing more water to evaporate out of the leaf.

Chapter 16
Animals and the Organ Systems

16.1   1. a   2. d   3. Stem cells are undifferentiated cells that are used to produce other, specialized offspring cells in an organism. As the offspring cells block off some genes inherited from the stem cell and express others, they take on a special structure and function and thus become specialized through the process of differentiation.

16.2   1. d   2. b   3. If kidney cells recognize that the water level has fallen below homeostatic levels, then hormones encourage the lining to become less porous so the kidney retains more water and less is lost in the excretory product. If water levels are high, then hormones encourage water to leave the kidney and form more dilute urine.

Chapter 17
Populations and Communities

17.1   1. d   2. a   3. Interspecific interactions like predator-prey and symbiosis illustrate interdependence in nature in that they involve two different species interacting in a unique way. As one organism in the interaction responds, it affects the function of the other species.

17.2   1. c   2. a   3. Density-dependent limiting factors represent an interdependence because they tend to be associated with the life processes of other living things. For example, the population levels of bacteria may have an effect on the level of parasitism that can occur within hosts. This in turn can affect the density and distribution of the host population.

17.3   1. d   2. d   3. Predator-prey interactions represent an interesting type of interdependence. While the predator obviously benefits from the prey that it consumes and the prey is harmed by the relationship, the prey population actually benefits from the interaction. The remaining prey tend to be healthier and more evolutionarily fit than the prey that have been consumed, and they are left to contribute to the gene pool used to repopulate future generations of prey.

Chapter 18
Ecosystems and Biomes

18.1   1. d   2. b   3. The Earth's carbon cycle is out of balance because of the great quantities of CO_2 that are released into the atmosphere from the unnatural burning of fossil fuels and the simultaneous deforestation within many ecosystems. Thus, fewer CO_2-fixing organisms are around to absorb the excess gas that is present.

18.2   1. b   2. c   3. Eutrophic lakes and slow-moving rivers both contain higher nutrient levels and thus support more life than oligotrophic lakes and fast-moving rivers, respectively. The higher nutrient levels give the water a murkier appearance than it would otherwise have.

Illustration and Photo Credits

Page 15: (a) Volker Brinkmann, Max Planck Institute for Infection Biology, Berlin, Germany. Published in *Public Library Science Journal*; (b) Y Tambe

Page 19: (top of page) Julian Habekost and Jonas Konrad

Page 25: Yikrazuul

Page 32: Bruce Wetzel, Harry Schaefer (photographers), National Cancer Institute

Page 39: National Human Genome Research Institute

Page 44: Content provided by Mariana Ruiz, LadyofHats

Page 45: Content provided by Mariana Ruiz, LadyofHats

Page 49: Daniel Mayer

Page 54: Based on work by Frank Boumphrey, M.D.

Page 60: David O. Morgan, from The Cell Cycle: Principles of Control, Science Press

Page 63: Content provided by Mariana Ruiz, LadyofHats

Page 70: Based on work by Frank Boumphrey, M.D.

Page 71: Ehamberg

Page 80: Based on work by Daniel Horspool

Page 83: Based on work by Madeleine Price Ball

Page 93: National Human Genome Research Institute

Page 97: Content provided by National Institutes of Health

Page 100: James Tourtellotte, United States Department of Homeland Security

Page 105: Content provided by Carmel830

Page 107: Based on work by Yassine Mrabet

Page 115: Cambridge University Library

Page 118: Vladislav Petrovich Volkov

Page 127: Based on work by Peter Halasz

Page 130: Based on work by Madeleine Price Ball

Page 138: Content provided by Mariana Ruiz, LadyofHats

Page 142: Kankitsurui

Page 145: Based on work by Debivort

Page 147: Kathy E. Goodness

Page 150: Peter Coxhead

Page 159: Fire Salamander by Christian Jansky; Saltwater Crocodile by Paul Thomsen; Ocean Sunfish by NOAA; Elephant Shrew by Joey Makalintal; Southern Dassowary by B. S. Thurner Hof

Page 166: Content provided by Mariana Ruiz, LadyofHats

Page 167: Luis Fernández García

Page 174: Content provided by Mariana Ruiz, LadyofHats

Page 175: Content provided by the National Institute of Diabetes and Digestive and Kidney Diseases, National Institutes of Health

Page 176: Content provided by the National Institute of Diabetes and Digestive and Kidney Diseases, National Institutes of Health

Page 182: Content provided by the United States Department of Agriculture

Page 184: Yerpo

Page 185: Based on work by Simon Pierre Barrette

Page 188: Jan Derk

Page 192: Wasserkreislauf

Page 194: Based on work by Mark David Thompson

Page 195: USDA Natural Resources Conservation Service

Index